EXPERIMENTAL PSYCHOLOGY

A Methodological Approach

FRANK J. McGUIGAN

Professor of Psychology   Hollins College

PRENTICE-HALL, INC.

Englewood Cliffs, N.J.

1 9 6 0

To two charming ladies

CONSTANCE AND JOAN

© 1960 by Prentice-Hall, Inc., Englewood Cliffs, N.J.

Library of Congress Catalog Card Number 60-14197

Printed in the United States of America

29502-C

Preface

Experimental psychology was born with the study of sensory processes; it grew as additional topics, such as perception, reaction time, attention, emotion, learning, and thinking, were added. Accordingly, the traditional course in experimental psychology was a course the content of which was accidentally defined by those lines of investigation followed by early experimenters in those fields. But times change, and so does experimental psychology. The present trend is to define experimental psychology not in terms of specific content areas, but rather as a study of scientific methodology generally, and of the methods of experimentation in particular. There is considerable evidence that this trend is gaining ground rapidly.

This book has been written to meet this trend. His methods no longer confined to but a few areas, the experimental psychologist conducts researches in almost the whole of psychology—clinical, industrial, social, military, and so on. To emphasize this point, we have throughout the book used examples of experiments from many fields, illustrative of many methodological points.

In short, then, the point of departure for this book is the relatively new conception of experimental psychology in terms of methodology, a conception which represents the bringing together of three somewhat distinct aspects of science: experimental methodology, statistics, and philosophy of science. We have attempted to perform a job analysis of experimental psychology, presenting the important techniques that the experimental psychologist uses in his everyday work. Experimental methods are the basis of experimental psychology, of course; the omnipresence of statistical presentations in journals attests the importance of this aspect of experimentation. An understanding of the philosophy of science is important to an understanding of what science is, how the scientific method is used, and particularly of where experimentation fits into the more general framework of scientific methodology. With an understanding of the goals and functions of scientific methodology, the experimental psychologist

iii

is prepared to function efficiently, avoiding scientifically unsound procedures and fruitless problems.

Designed as it is to be practical in the sense of presenting information on those techniques actually used by the working experimental psychologist, it is hoped for this book that it will help maximize transference of performance from a course in experimental psychology to the type of behavior manifested by the professional experimental psychologist.

Acknowledgment

I cannot adequately express the appreciation and indebtedness that I feel to all those persons who have helped me, in a wide variety of ways, with this undertaking. Hollins College has been most generous in furnishing an atmosphere conducive to academic endeavor. My students have furnished both valuable criticism of ideas and exposition, and the reinforcement required for the completion of this project. Among those, however, who offered specific suggestions I find that I am particularly indebted to Drs. Allen Calvin, Victor Denenberg, David Duncan, Paul Meehl, Michael Scriven, Kenneth Spence, Lowell Wine, and Mr. John Berserth. I am also appreciative of the work of Charlotte Fisher and Blanche Buterbaugh for typing a readable manuscript out of a series of near-illegible notes.

I am indebted to the various authors and publishers who so kindly permitted me to draw on their sources (as acknowledged in the text), including Professor Sir Ronald A. Fisher, Cambridge, and to Messrs. Oliver and Boyd Ltd., Edinburgh, for permission to reprint pages Nos. 91-92 from their book *The Design of Experiments;* to reprint Table No. V from their book *Statistical Tables of Biological, Agricultural, and Medical Research;* and to reprint Table No. IV from their book *Statistical Methods for Research Workers.*

F.J.M.

Contents

I

An Overview of
Experimentation

The Nature of Science

One of the main differences between humans and the lower animals is man's ability to engage in abstract thinking. For instance, man is much more able to survey a number of diverse items and abstract certain characteristics that they have in common. In attempting to arrive at a general definition of science we might well proceed in such a manner. That is, we might consider the various sciences as a group and abstract the salient characteristics that distinguish them from other disciplines. Figure 1.1 is a schematic representation of the disciplines which man studies, rather crudely categorized into three groups (excluding the formal disciplines, mathematics and logic). Within the inner circle we have represented what are commonly called the sciences. The next circle embraces various disciplines that are not usually thought of as sciences, such as the arts and some of the humanities. Outside of that circle are yet other disciplines which, for lack of a better term, are designated as metaphysical disciplines.

The sciences in the inner circle certainly differ among themselves in a number of ways. But in what important ways are they similar to each other? Likewise, what are the similarities among the disciplines in the outer circle? What do the metaphysical disciplines outside the circle have in common? Furthermore, in what important ways do each of these three groups differ from each other? Answers to these questions should enable us to arrive at an approximation to a general definition of "science."

One common characteristic of the sciences is that they all use the same general approach in solving problems—"the scientific method." The "scientific method" is a serial process by which all the sciences obtain answers to their problems. Neither of the other two groups explicitly uses this method.

1

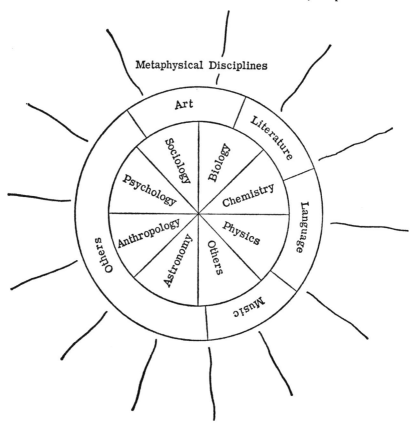

Figure 1.1. Three groups of disciplines which man studies. Within the inner circle are the sciences. The second circle contains the arts and some types of the humanities, while the metaphysical disciplines fall outside the circles.

We may also note that the disciplines within the two inner circles differ from the metaphysical disciplines in type of problem studied. Those disciplines within the two circles attempt to study "meaningful" problems, while those outside of the circle generally study "meaningless" problems. Briefly, a meaningful problem is one that can be answered with the use of man's normal capacities. A meaningless problem is one that is essentially unanswerable. Such meaningless problems usually concern supernatural phenomena or questions about *ultimate* causes. For example, the question of what caused the universe is meaningless and typical of studies in religion, classical philosophy, etc.[1] The problem

[1] Crude categorizations are dangerous. We want to point out *general* differences among the various disciplines. A number of theological problems, for example, are meaningful; however most theologians are not interested in answering such questions (e.g., does praying affect our everyday life) in an empirical manner.

of meaning is an extremely important one and will be taken up in greater detail in Chapter 2.

"Meaningful" and "meaningless" are technical terms and certain vernacular meanings should not be read into them. It is not meant, for instance, to establish a hierarchy of values among the various disciplines by classifying them according to the type of problem they study. We are not necessarily saying, for example, that the problems of science are "better" than the problems of religion. The distinction is that meaningful problems are capable of being solved, but meaningless ones are not. The problems of the metaphysical disciplines are in no sense of any lesser importance—in fact, many intelligent men would argue that they are more important. Those who stay within the two circles simply believe they must limit their study to problems that they are capable of solving. Of course, many scientists also devote part of their lives to the consideration of supernatural phenomena. But it is important to realize that when they do, they have "left their circle" and are, for that time, no longer behaving as scientists.

In summary, first, the sciences use the scientific method, and they study meaningful problems. Second, the disciplines in the outer circle do not use the scientific method, but they do study meaningful problems. And third, the disciplines outside of the circles neither use the scientific method nor study meaningful problems. These considerations lead to the following definition—*"Science" is the application of the scientific method to meaningful problems.* Generally, neither of the other two groups of disciplines have *both* these features in common.[2]

With this general definition in hand let us consider the scientific method, primarily as it is applied in psychology. And since the most powerful application of the scientific method is experimentation, we shall focus primarily on how experiments are conducted. The problems with which psychologists are concerned are among the most challenging and complex that man faces. For this reason it is necessary to bring to bear the most effective methods that science can make available in attempting to solve them. The following brief discussion will serve as an overview of the rest of the book. By studying a general picture of how the experimental psychologist proceeds, you should be able to obtain a general orientation to experimentation. Because this overview is so brief, however, complex matters will necessarily be oversimplified. Possible distortions resulting from this oversimplification will be corrected in later chapters.

[2] It is likely that there is no completely adequate definition of science available. We emphasize that there are limitations to this one, but that an understanding of it will facilitate presentation of later material.

Psychological Experimentation—An Application of the Scientific Method[3]

A psychological experiment starts with the formulation of a problem, which is usually best stated in the form of a question. The only requirement that the problem must meet is that it be meaningful—it must be answerable with the tools that are available to the psychologist. Beyond this, the problem may be concerned with any aspect of behavior, whether it is judged to be important or trivial. One lesson of history is that we must not be hasty in judging the importance of the problem on which a scientist works, for many times what was discarded as being of little importance contributed sizeably to later scientific advances.

The experimenter generally expresses a tentative solution to the problem. This tentative solution is called a *hypothesis*, and it may be a reasoned potential solution or it may be only a vague guess (it is an *empirical* hypothesis, not a null hypothesis, which will be discussed in Chapter 5.) Following the statement of his hypothesis, the experimenter seeks to determine whether the hypothesis is probably true or probably false, i.e., does it answer the problem he has set for himself? To answer this question he must collect data, for a set of data is his only criterion. Various techniques are available for data collection, but we are mainly concerned with the use of experimentation for this purpose.

One of the first steps that the experimenter will take in actually collecting his data is to select a group of subjects with which to work. The type of subject he studies will be determined in part by the nature of the problem. If he is concerned with psychotherapy, he may select a group of mentally disturbed patients. A problem concerned with the function of parts of the brain would entail the use of animals (for few humans volunteer to serve as subjects for brain operations). Learning problems may be investigated with the use of college sophomores, chimpanzees, rats, etc. But whatever the type of subject, the experimenter will assign them to groups. We shall consider here the basic type of experiment, namely one that only involves two groups.

The assignment of subjects to groups must be made in such a way that the groups will be approximately equivalent at the start of the experiment. The experimenter next typically administers an experimental treatment to one of the groups. The experimental treatment is what he wishes to evaluate, and it is administered to the *experimental group*. The other

---

[3] There are those who hold that psychologists do not formally go through the following steps of the scientific method in conducting their research. We would agree with this statement for many researchers. However, a close analysis of the actual work of such people would suggest that they at least informally approximate the following pattern.

group, called the *control group*, usually receives a normal or standard treatment. It is important, here, to understand clearly just what the terms "experimental" and "normal" or "standard treatment" mean.

In his study of behavior, the psychologist generally seeks to establish empirical relationships between aspects of the environment, broadly conceived, and aspects of behavior. These relationships are known by a variety of names, such as hypotheses, theories, or laws. Such relationships in psychology essentially state that if a certain environmental characteristic is changed, behavior of a certain type also changes.[4]

The aspect of the environment which is experimentally studied is called the *independent variable;* the resulting change in behavior is called the *dependent variable.* Essentially, a variable is anything that changes in value. It is a quality that can exhibit differences in value, usually in magnitude or strength. Thus it may be said that a variable generally is anything that may assume different numerical values. Psychological variables change in value from time to time for any given organism, between organisms, and according to various environmental conditions. Some examples of variables are the height of men, the weight of men, the speed with which a rat runs a maze, the number of trials required to learn a poem, the brightness of a light, the number of words a patient says in a psychotherapeutic interview, and the amount of pay a worker receives for performing a given task.

Figure 1.2 schematically represents one of these examples, "the speed

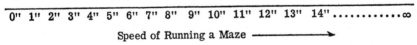

0" 1" 2" 3" 4" 5" 6" 7" 8" 9" 10" 11" 12" 13" 14" . . . . ∞

Speed of Running a Maze ⟶

Figure 1.2. Diagramatic representation of a continuous variable.

with which a rat runs a maze." It can be seen that this variable can take on any of a large number of magnitudes, or more specifically, it can exhibit any of a large number of time values. In fact, it may "theoretically" assume any of an infinite number of such values, the least being zero seconds, and the greatest being an infinitely large amount of time. In actual situations, however, we would expect it to exhibit a value of a

[4] By saying that the psychologist seeks to establish relationships between environmental characteristics and aspects of behavior, we are being unduly narrow. Actually he is also concerned with unobservable processes (variously called logical constructs, intervening variable, hypothetical constructs, etc.). Since, however, it is unlikely that your elementary work will involve hypotheses of such an abstract nature, they will not be further discussed. The highly arbitrary character of defining and differentiating among the various kinds of relationship should be emphasized—frequently the grossly empirical kind of relationship that we are considering under the label "hypothesis," once it is confirmed, is referred to as an empirical or observational law, or before it is tested merely as "hunch" or "guess."

number of seconds, or at the most, several minutes. But the point is that there is no limit to the specific time value that it may assume, for this variable may be expressed in terms of any number of seconds, minutes, hours, etc., including any fraction of these units.

For example, we may find that a rat ran a maze in 24 seconds, in 12.5 seconds, or in 2 minutes, 19.3 seconds. Since this variable may assume any fraction of a value (it may be represented by any point along the line in Figure 1.2), it is called a *continuous* variable. A continuous variable is one that is capable of changing by any amount, even an infinitesimally small one. A variable that is not continuous is called a *discontinuous* or *discrete* variable. A discrete variable can assume only numerical values that differ by clearly defined steps with no intermittent values possible. For example, the number of people in a theatre would be a discrete variable, for, barring an unusually messy affair, one would not expect to find a part of a person in such surroundings. Thus, one might find 1, 15, 299 or 302 people in a theatre, but not 1.6 or 14.8 people. Similarly, sex (male or female), eye color (brown, blue, etc.) are frequently cited as examples of discrete variables.[5]

We have said that the psychologist seeks to find relationships between independent and dependent variables. There are an infinite (or at least indefinitely large) number of independent variables available in nature for the psychologist to examine. But he is interested in discovering those *relatively* few ones that affect a given kind of behavior. In short, we may say that an independent variable is any variable that is investigated for the purpose of determining whether it influences behavior. Some independent variables that have been investigated in experiments are age, hereditary factors, endocrine secretions, brain lesions, drugs, race, and home environments.

To determine whether a given independent variable affects behavior the experimenter administers one value of it to his experimental group and a second value of it to his control group. The value administered to the experimental group is as we have said, the "experimental treatment," while the control group is usually given the "normal treatment." Thus, the essential difference between the "experimental" and "normal" treatment is the specific value of the independent variable that is assigned to each group. For example, the independent variable may be the intensity of a light (a continuous variable). The experimenter may subject the

---

[5] A more advanced consideration of discrete variables cannot be offered here. We may simply note that some scientists question whether there are actually any discrete variables in nature. They suggest that we simply "force" nature into "artificial" categories. Color, for example, may more properly be conceived of as a continuous variable—there are many gradations of brown, blue, etc. Nevertheless, scientists find it useful to categorize variables into classes as discrete variables, and to view such categorization as an approximation.

experimental group to a high intensity, and the control group to a zero intensity.

For a better understanding of the nature of an independent variable, let us consider another example of how one might be used in an experiment. Visualize a continuum similar to Figure 1.2, composed of an infinite number of possible values that the independent variable may take. If, for example, we are interested in determining how well a task is retained as a result of the number of times it is practiced, our continuum would start with zero trials and continue with one, two, three, etc. trials (this would be a discrete variable).

Let us suppose that in a certain industry workers are trained by performing an assembly line task 10 times before being put to work. After a time, however, it is found that the workers are not assembling their product adequately and it is judged that they have not learned their task sufficiently well. Some corrective action is indicated and the foreman suggests that the workers would learn the task better if they were able to practice it fifteen times instead of ten. Here we have the makings of an experiment of the simplest sort.

We may think of our independent variable as the "number of times that the task is performed in training," and will assign it two of the possibly infinite number of values that it may assume—ten trials and fifteen trials. (See Figure 1.3.) Of course, we could have selected any

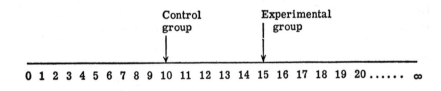

Figure 1.3. Diagramatic representation of an independent variable as a continuum. The value of the independent variable assigned to the control group is 10 trials; that assigned to the experimental group, 15 trials.

number of other values, 1 trial, 5 trials, or five thousand trials, but because of the nature of the problem with which we are concerned 10 and 15 seem the best values to study. We will have the experimental group practice the task fifteen times, the control group ten times. Thus, the control group receives the normal treatment (ten trials), and the experimental group is assigned the experimental or new treatment (fifteen trials). In many cases, of course, it is arbitrary which group is labeled the control group and which is called the experimental group. Sometimes both treatments are novel ones, in which case it is impossible to label the groups in this manner—they might simply be called "Group 1" and "Group 2." In another instance, if one group is administered a "zero" value of the independent variable, and a second group is given some positive amount of that variable, then the zero treatment group would be called the "control group" while the other would be the "experimental group."

The *dependent* variable is usually some well defined aspect of behavior (a response) which the experimenter measures in his experiment. It may be the number of times the subject says a certain word, the rapidity with which a subject learns a given task, the number of items a worker on a production line can turn out in an hour, and so on. The value obtained for the dependent variable is the criterion of whether or not the independent variable is effective. It is in this sense that it is called a *dependent* variable—the value that it assumes is expected to be dependent on the value assigned to the independent variable.[6] Thus, an experimenter will vary the independent variable and note whether the dependent variable changes. If the dependent variable changes in value as the independent variable is manipulated, then it may be asserted that there is a relationship between the two. If the dependent variable does not change, however, it may be asserted that there is a lack of relationship between them. For example, let us assume that a light of high intensity is flashed into the eyes of each subject of the experimental group, while those of the control group are not subjected to any light. The dependent variable might be the amount of contraction of the iris diaphragm (the pupil) of the eye which, it may be noted, is an aspect of behavior—a response. If we find that the average contraction of the pupil of the experimental subjects exceeds that of those in the control group, we may conclude that intensity of light is an effective independent variable. We can assert the following relation: the greater the intensity of a light that is flashed into a subject's eyes, the greater the contraction of the pupil. If, on the other hand, we find no difference in the amount of contraction of the pupil between the two groups, we would assert that there is a lack of relationship between these two variables.

Perhaps the most important principle of experimentation, stated in an ideal form, is that the experimenter must hold constant all of the variables that may affect his dependent variable, except the independent variable(s) that he is attempting to evaluate.[7] Obviously, there are a number of variables that may affect the dependent variable, but the experimenter is not immediately interested in these. He is, for the moment,

[6] Assuming, of course, that the independent variable is effective. It may be added that the dependent variable is also dependent on some of the extraneous variables, discussed later, that are present in the experiment.

[7] Further reading will show why this is an ideal statement, and the ways in which it needs to be modified for any given experiment.

interested in only one thing—the relationship, or lack of it, between his independent and his dependent variable. If the experimenter allows a number of other variables to operate in the experimental situation (call them *extraneous* variables) his experiment is going to be contaminated. For this reason he must *control* the important extraneous variables in his experiment. There are various techniques available for controlling extraneous variables, many of which are discussed in Chapter 6.

A simple illustration of how an extraneous variable might contaminate an experiment and thus make the findings unacceptable might be made using the last example. Suppose that the intensity of light is varied on subjects who have just received a vaccination. But the serum contains a substance that affects the contraction of the pupil of the eye. The data obtained will obviously be useless. If the effect was such as to cause the pupil not to contract, the experimental subjects would show about the same amount of contraction (or rather, lack of contraction) as the control subjects. We would thus conclude that the independent variable did not affect the response being studied. Our findings would assert that these two variables of light and pupil contraction are not related, when in fact they are.

With this discussion of independent and dependent variables behind us let us return to our general discussion of the scientific method as applied to experimentation. We have said that a scientist starts his investigation with the statement of a problem, after which he advances a hypothesis as a tentative solution to that problem. He may then conduct an experiment to collect data, data which should indicate the probability that his hypothesis is true or false. He may find it advantageous to use certain types of apparatus and equipment in his experiment. The particular type of apparatus used will naturally depend on the nature of the problem. In general, apparatus is used in an experiment for two main reasons. First, to administer the experimental treatment, and second, to allow, or to facilitate, the collection of data.

The hypothesis that is being tested will predict the way in which the data should point. It may be that the hypothesis will predict that the experimental group will perform better than the control group. By confronting the hypothesis with the dependent variable scores of the two groups the experimenter can determine if this is so. But it is difficult to tell whether the (dependent variable) scores for one group are higher or lower than the scores for the second group simply by looking at a number of unorganized data for the two groups of subjects. Therefore, the experimenter must reduce all of the data with which he is dealing to numbers that can be reasonably handled, numbers that will provide him with an answer— For this reason, he must resort to statistics.

For example, he may compute an average (mean) score for both the experimental group and the control group. He might find that the experimental group has a higher mean score (say, 100) than the control group (say, 99). While we note that the experimental group has a higher mean score, we also note that the difference between the two groups is very small. Is this difference, then, a "real" difference, or is it only a chance difference. What are the odds that if we conducted the experiment again we would obtain the same results? If the difference is a "real," reliable difference, then the experimental group should obtain a higher mean score than the control group almost every time the experiment is repeated. But if there is no "real" difference between the two groups we would expect to find the experimental group receiving the higher score half of the time, and the control group being superior the other half of the time. To tell whether the results of one experiment are "real," rather than simply due to chance, the experimenter resorts to any of a variety of statistical tests. The particular statistical test(s) that he uses will be determined by the type of data obtained and the general design of the experiment. But the point is that, on the basis of such tests, it can be determined whether the difference between the two groups is "real" and reliable or merely "accidental." More appropriately, the tests indicate whether or not the difference is statistically *significant,* for this is what is meant by "real" and "reliable" differences. If the difference between the dependent variable scores of the groups is significant, it may be assumed that this is not due to chance and that the independent variable is effective.

Thus, by starting with two equivalent groups, administering the experimental treatment to one but not to the other, and collecting and analyzing the data thus obtained, suppose we find a significant difference between the two groups. We may legitimately assume that the two groups eventually differed because of the experimental treatment. Since this is the result that was predicted by our hypothesis, the hypothesis is supported, or confirmed. In other words, when a hypothesis is supported by experimental data, that hypothesis is probably true. On the other hand, if, in the above example, the control group is found to be equal or superior to the experimental group, the hypothesis is not supported by the data and we may conclude that it is probably false. Naturally, this step of the scientific method in which the hypothesis is tested is considerably oversimplified in our brief presentation. It will be necessary to consider the matter more thoroughly later (Chapter 11.)

Closely allied with testing of the hypothesis is an additional step of the scientific method, "generalization." After completing the phases outlined above, the experimenter may feel quite confident that the hypothesis is true for the specific conditions under which he tested it. He must underline *specific* conditions, however, and not lose sight of how specific they are in any given experiment. But the work of the scientist *qua* scientist is not concerned with truth under any *specific* set of conditions. Rather, he usually wants to make as *general* a statement as he possibly can about nature. And herein lies much of his joy and grief, for the more he generalizes his findings, the greater are the chances for error. Suppose that he has used college students as the subjects for his experiment. This selection does not mean that he is interested *only* in the behavior of college students. Rather, he is probably interested in the behavior of *all* human beings, and perhaps even of *all* organisms. Because he has found his hypothesis to be probably true for his particular group of subjects, can he now say that it is probably true for all humans? Or must he simply restrict his results to college students? Or must he narrow the focus even further, limiting it to those students attending the college at which he conducted his experiment. This, essentially, is the question of generalization—how widely can the experimenter generalize his results? He wants to generalize as widely as possible, yet not so widely that the hypothesis "breaks down." The question of how widely he may safely generalize his hypothesis will be discussed at greater length in Chapter 13. The broad principle to remember is that he should state that his hypothesis is applicable to as wide a set of conditions (e.g., to as many classes of subjects) as the nature of his experiment warrants.

The next step in the scientific method, closely related to the preceding ones, concerns making predictions on the basis of the hypothesis. By this we mean that a hypothesis may be used to predict certain events in new situations—to predict for example, that a different group of subjects will act in the same way as a group studied in an earlier experiment.[8] We can add a final step in the scientific method, *replication*. In *replication* the experimenter conducts an additional experiment. He uses the confirmed hypothesis as the basis for predicting that a new sample of subjects will behave as did the original sample. If the prediction made by the use of the previously confirmed hypothesis is found to hold in the new situation, the probability that the hypothesis is true is tremendously increased.

In summary let us set down the various steps in the scientific method. (Be advised, however, that there are no rigid rules to follow in doing this. In any process that one seeks to classify into a number of arbitrary categories, some distortion is inevitable. Another source might offer a different classification, while still another one might refuse, quite legitimately, to attempt such an endeavor.)

First, the scientist states a problem that he wishes to investigate. Next, he formulates the hypothesis, a tentative solution to that problem. Third,

[8] A case can be made for not including prediction as a part of the scientific method, at least in some sciences (cf. Scriven, 1959).

he collects data relevant to the hypothesis. Following this, he tests the hypothesis by confronting it with the data and makes the appropriate inferences—he organizes the data through statistical means and determines whether the data support or refute the hypothesis. Fifth, assuming that the hypothesis is supported, he may wish to generalize to all things with which the hypothesis is legitimately concerned, in which case he should explicitly state the generality with which he wishes to advance his hypothesis. Sixth, he may wish to make a prediction to new situations, to events not studied in the original experiment. And finally, he may wish to test the hypothesis anew in the novel situation; that is, he might replicate (conduct a new experiment) to attempt to increase the probability of his hypothesis.

An Example of a Psychological Experiment

To make the preceding discussion more concrete, consider an example of how an experiment might be conducted from its inception to its conclusion. Let us assume that a psychotherapist has some serious questions about how best to proceed with his clients in order to affect a "cure" as efficiently as possible. It usually happens that psychotherapists become aware of the basic ("real") problems of their clients before the clients themselves do. Thus, they are generally in a position to offer advice to the clients. In our example, however, the psychotherapist is not sure whether offering direct advice is a good procedure to follow. Not only may the client ignore the advice, but he may even react violently against it, thus retarding the therapeutic process. The problem may be stated as follows: Should a psychotherapist authoritatively advise his clients what their problems are and what they should do about them, or should he just sit back and let the clients arrive at their own assessment of the problem and determine for themselves the best course to take? Assume that the psychotherapist believes the latter to be the better procedure. The reasons for or against his opinion need not detain us here. We simply note his hypothesis: If a client undergoing psychotherapy is allowed to arrive at the determination of his problem and its proposed solution by himself, then his recovery will be more efficient than if the psychotherapist gives him this information in an authoritative manner. We might identify the independent variable as "the amount of guidance furnished to the clients," and assign two values to it: first, a maximal amount of guidance, and second, a zero (or at least minimal) amount of guidance.[9]

Suppose that the psychotherapist has ten clients, and that he assigns them to two groups of five each. A great deal of guidance will then be

---

[9] This example well illustrates that *frequently* it is not appropriate to say that a zero amount of the independent variable can be administered to a control group.

given to one of the groups, while a minimum amount will be administered to the second group. The group that receives a minimum amount of guidance will be called the control group; the group that receives the maximum amount of guidance will be called the experimental group.

Throughout the course of therapy, then, the psychotherapist administers the two different treatments to the experimental and control groups. During this time he prevents the important extraneous variables from acting differently on the two groups. For example, he would want the clients from both groups to undergo therapy in the same place (his office, for instance) to eliminate the possibility that the progress of one group might differ from that of the other group because of the immediate surroundings in which the therapy takes place.

The dependent variable here may be specified as the progress toward recovery. Such a variable is obviously rather difficult to measure, but for illustrative purposes we might use a time measure. Thus, we might assume that the earlier the client is discharged by the therapist, the greater is his progress toward recovery. Assuming that the extraneous variables have been adequately controlled, the progress toward recovery (the dependent variable) depends on the particular values of the independent variable used, and on nothing else.

As therapy progresses the psychotherapist collects his data. Specifically, he determines the amount of time each client spends in therapy before he is discharged. After all of the clients are discharged, the therapist compares the times for the experimental group against those for the control group. Let us assume that he finds that the mean amount of time in therapy of the experimental group is higher than that of the control group, and further, that a statistical test indicates that the difference is significant. That is, the group that received minimum guidance had a significantly lower time-in-therapy (the dependent variable) than did the group that received maximal guidance. It will be recalled that this is precisely what the therapist's hypothesis predicted. Since the results of the experiment are in accord with the hypothesis, we may conclude that the hypothesis is confirmed.

Now the psychotherapist is happy, since he has solved his problem and now knows which of the two methods of psychotherapy is better. But has he found "truth" only for himself, or is what he has found applicable to other situations—can other therapists also benefit by his results? Can his findings be extended, or generalized, to all therapeutic situations of the nature that he has studied? After serious consideration, he decides to assert that his findings are applicable to the psychotherapy conducted by other psychologists and publishes his findings in a scientific journal.

Inherent in the process of generalization is that of prediction (although there can be generalizations that are not used to make predictions). Here, in effect, what the therapist is doing when he generalizes, is predicting—predicting that the same results would be obtained if a new situation were studied. In this simple case the therapist is essentially saying that for other subjects, offering minimal guidance would result in more rapid recovery than if maximal guidance is offered them. To test this prediction, another psychotherapist conducts an experiment as outlined above (the experiment is replicated). His findings prove to be the same, and the hypothesis is again supported by the data. With this independent confirmation of the hypothesis as an added factor, it may be concluded that the probability of the hypothesis is increased. That is, our confidence that the hypothesis is true is considerably greater than before.[10]

[10] The oversimplication of several topics in this chapter is especially apparent in this fictitious experiment. For instance, the important extraneous variable of the therapist's own confidence in one method of psychotherapy is undoubtedly going to affect the results, and we have left this variable uncontrolled.

2

The Problem

What is a Problem?

A contemporary scientific inquiry starts with a problem; or more accurately, when sufficient knowledge has already been collected to indicate that a problem exists. But this knowledge is not sufficient to offer a solution, for at least two reasons. There may not be enough of it to answer the problem; or, what knowledge we do possess may be in such a state of disorder that it cannot be adequately related to the problem. The latter reason requires an elaboration that is best left to a later discussion. The former, however, can be profitably discussed here.

Ways in Which a Problem Is Manifested

The lack of a sufficient quantity of knowledge that bears on a problem is manifested in at least three, to some-extent-overlapping, ways: first, when the results of several inquiries disagree; second, when there is a noticeable gap in the results of investigations; and third, when a "fact" exists in the form of a bit of unexplained information. Let us consider each of these in greater detail.

Contradictory Results

To understand how the results of different attempts to solve the same problem may differ, consider three separate experiments that have been reported in psychological journals. All three experiments were very similar in nature, and they all addressed themselves to the following problem: "When a person is learning a task, are rest pauses more beneficial if concentrated during the first part of the total practice session or if concentrated during the last part?" For instance, if a person is to spend ten trials in practicing a given task, would his learning be more efficient if his rest pauses were concentrated between the first five trials (early in learning), or between the last five trials (late in learning)? The general design of all three experiments was as follows: One group of subjects practiced a task with rest pauses concentrated during the early part of the practice session. As these subjects continued to practice the task on additional trials, the length of the rest pauses between trials progressively decreased. A second group of subjects practiced the task with progressively increasing rest pauses between trials—as the number of trials on which they practiced the task increased, the amount of rest between trials became larger.

The first experiment indicated that progressively increasing rest periods are superior (Doré and Hilgard, 1938), the second experiment showed that progressively decreasing rest periods led to superior learning (Renshaw and Schwarzbeck, 1938), while the third experiment indicated that the effects of progressively increasing and progressively decreasing rest periods are about the same (Cook and Hilgard, 1949). Why do these three studies provide us with conflicting results?

One possible reason for conflicting results is that one or more of the experiments was poorly conducted—certain principles of sound experimentation may have been violated. Perhaps the most common error in experimentation is the failure to control important extraneous variables. To demonstrate briefly how such a failure may produce conflicting results, let us assume that one important extraneous variable is not considered by the experimenter. Unknown to the experimenter, this variable is actually influencing the dependent variable. In one experiment it happens to assume one value, while in a second experiment on the same problem it happens to assume a different value. Thus, it leads to different values for the dependent variable in the two experiments. There are many other reasons why the results of an experiment may be confusing. Let us simply note now that the problem in this case is: "Why do we have conflicting results?" With additional knowledge of the extraneous variables involved and of what we shall later discuss as the *interaction* among variables, we may be able to ascertain the reason for conflicting results.

A Gap in Our Knowledge

The second way in which we become aware of a problem is when there is a noticeable gap in the results of investigations. Perhaps a problem is most easily manifested in this way—we are simply aware of what we do know and that there is something which we do not know. If a community group plans to establish a clinic to provide psychotherapeutic services, two natural questions for them to ask are, "What kind of psychotherapy should we offer?" and "Of the different systems of psychotherapy, which is the most effective?" Now these questions are extremely important, but there are few scientifically acceptable studies that provide answers. Here is an apparent gap in our knowledge. Collection of data with a view toward filling this gap is thus indicated.

Explaining a Fact

The third way in which we become aware of a problem is when we are in possession of a "fact," and we ask ourselves "Why is this so?" A fact, existing in isolation from the rest of our knowledge, demands *explanation*.

Science consists not only of knowledge, but of *systematized* knowledge. The greater the systematization, the greater is the scientist's understanding of nature. Thus, when a new fact is acquired, the scientist seeks to relate it to the already existing body of knowledge. But he does not know exactly where in his framework of knowledge the new fact fits, or even that it will fit. If, after sufficient reflection, he is able to relate the new fact adequately to the existing knowledge, it may be said that he has *explained* it. That fact presents no further problem. On the other hand, if the fact does not fit in with existing knowledge, a problem is made apparent. The collection of additional information is necessary so that eventually, the scientist hopes, the new fact will be related to additional information in such a manner that it will be "explained." By this gradual process, the scientist's understanding and control of nature is extended. Some problems of this kind will lead to little that is of significance for science, while others may result in major discoveries. Examples of new portions of knowledge that have had revolutionary significance are rare in psychology since it is such a new science, but they are relatively frequent in other sciences.[1] To illustrate how the discovery of a new fact has created a problem the solution of which has had important consequences, consider an example.

One day the Frenchman, Henri Becquerel found that a piece of photographic film had been fogged. He could not immediately explain this, but in thinking about it he noticed that a piece of uranium had been placed near the film before the fogging. Existing theory did not suggest that there was any connection between the uranium and the fogged film. But Becquerel sought to explain this fact by suggesting that the two events were connected to each other. To relate the events more specifically, he had to postulate that the uranium gave off some unique kind of energy. Working along these lines, eventually he determined that the metal gave off radioactive energy which caused the fogging. This discovery led to a whole series of developments which have resulted in present-day theories of radioactivity.

[1] Wertheimer's attempts to explain the phi phenomenon may be one such case in psychology.

One of two psychological reactions often occurs to people when they discover a bit of new information. The curious, creative person will adventurously attempt to explain it. The incurious and unimaginative person, on the other hand, may attempt to ignore the problem, hoping it will "go away." A good example of the latter type of reaction occurred around the 15th century. Mathematicians had produced a "new" number that they called "zero." The thought that zero could be a number was disturbing to a number of people of that day. A number of city legislative bodies even passed laws forbidding the use of zero. The creation of imaginary numbers led to similar reactions; in some cases the entire arabic system of numerals was outlawed. Fortunately, this legislation was not effective. An interesting implication of the considerations presented above has been noted by Homer Dubs (1930): as our knowledge increases, so does the number of our problems. From our everyday observations of science this would seem to be true, for the answer to one problem often suggests other problems. Thus, the more our science advances, the more problems we will face.

A Problem Must Be Meaningful

Not all questions that arise can be answered by science. As noted in Chapter 1, a problem can qualify as the object of scientific inquiry only if it is meaningful, as distinguished from a meaningless, problem: science deals only with hypotheses that are *testable*. But since the word "meaningful" has been more widely used in the past, we shall continue to use it here. One of the most important activities related to science, and also one of the most complex, is the determination of a criterion of meaning. Such a criterion should enable us to determine whether or not a problem is meaningful. Because of the numerous discordant views on meaning it would be impossible to offer a presentation that would satisfy all. As a result we shall get into the problem of meaning only insofar as it should enter the everyday life of the experimental psychologist.[2]

The Truth Theory of Meaning

Briefly, we shall say that a problem is meaningful if it can be answered in a "yes" or "no" fashion. There are two important stages in the development of the theory of meaning that we are following here. The first stage is the statement of a truth or verifiability theory of meaning. The second

[2] For more advanced treatments of the topic refer to Reichenbach (1938), Frank (1956), and Feigl and Scriven (1956), especially the chapter in the latter by Rudolph Carnap entitled "The Methodological Character of Theoretical Concepts."

stage, an improvement on the truth theory of meaning, is the statement of a probability theory of meaning.

The main principle of the truth theory of meaning with which we shall be concerned may be expressed as follows: *A proposition is meaningful if, and only if, it is possible to determine that the proposition is either true or false.* It follows that only a proposition (a statement or sentence) can have meaning. Hypotheses are propositions, so we can use the truth theory of meaning to determine whether or not hypotheses are meaningful. For if it is possible to determine that a hypothesis is true or false, then the hypothesis is meaningful. But if it is not possible to determine that the proposition is either true or false, then the hypothesis is meaningless and should be discarded as being worthless to science.

Second, it follows that knowledge can only be expressed in the form of propositions. Problems are best stated in the form of questions, and such questions must be meaningful if they are to be subjected to scientific inquiry. When we say that a problem (stated as a question) is meaningful, we do not require that the problem be true or false. Rather, it must be possible to state a hypothesis as a potential answer to our problem, and it must be possible to determine that the hypothesis is either true or false. In short, a meaningful problem is one for which a meaningful hypothesis can be stated.[3]

The Probability Theory of Meaning

The words true and false have been used frequently in the above discussion. Strictly speaking, these words have been used only as approximations, for it is impossible to determine beyond all doubt that any given empirical proposition is true or false. The kind of world that we have for study is simply not of this nature. The best that we can do is to say that a certain proposition has a determinable degree of probability.[4] Thus, we cannot say in a strict sense that a certain proposition is true, but the best that we can say is that it is probably true. Similarly, we cannot say that another proposition is false; rather, we must say that it is highly improbable. Strictly speaking, then, the truth theory of meaning is inadequate for our purposes, for according to it no empirical proposition would ever be known to be meaningful since no empirical proposition can

[3] Of course the hypothesis must be relevant to the problem. For instance, if our problem is "What is the average height of Pigmies?" an irrelevant (but probably true) hypothesis would be "If a person smokes opium, then he will develop a dreamlike state." By relevance we shall mean that an inference can be made from the hypothesis to the problem, and the results of that inference constitute a solution to the problem.

[4] By "degree of probability" we mean that the proposition is true with a probability somewhere between 0 (absolutely false) and 1 (absolutely true). For instance, if a proposition has a probability of .5, then it is just as likely to be true as false.

ever be (absolutely) true or false. Hence we shall substitute the probability theory of meaning for the truth theory of meaning, the essential difference between the two being that the words "a degree of probability" are substituted for "true" and "false." The main principle of the probability theory of meaning with which we shall be concerned is: *A proposition is meaningful if, and only if, it is possible to determine a degree of probability for it.*

When we say that a proposition is meaningful, then, we understand it is meaningful as defined by the probability theory of meaning; i.e., that a proposition has meaning if it is possible to determine a degree of probability for it.

Kinds of Possibilities

Let us reconsider the probability theory of meaning. In particular, let us focus on the word "possible" contained in our statement of it. What does "possible" mean, *now*, or sometime in the future? Consider the question "Is it possible for man to fly to Uranus?" If by "possible" here we mean that one can step into a rocket ship today and set out on a successful journey, then clearly such a venture is not possible. But if we mean that such a trip is likely to be possible sometime in the future, then the answer is in the affirmative. In the following simplified discussion we shall consider two interpretations of "possible." The first interpretation we shall call *presently attainable*, and the second *potentially attainable*.

1. *Presently attainable.* This interpretation of "possible" concerns those possibilities that lie within the powers of people at the present time. If a certain task can be accomplished with the equipment that is immediately available we would say that the solution to the task is presently attainable. But if the task cannot be accomplished with tools that are presently available, the solution is not presently attainable. For example, building a bridge over the Suwannee River is presently attainable, but living successfully on the Sun is not presently attainable.

2. *Potentially attainable.* This interpretation concerns those possibilities that *may* come within the powers of people at some future time, but which are not possessed at the present. Whether or not they actually will be possessed in the future is a difficult matter to decide now. In the event that technological advances are sufficiently successful that the powers actually become possessed, then the potentially attainable becomes presently attainable. If, however, the immediate accomplishment of a certain task does not eventually become realizable, then it remains only potentially attainable, not being capable of being transferred to the category of presently attainable. For example, while a trip to Uranus is not presently attainable, we fully expect such a trip to be technologically feasible

in the future—successful accomplishment of such a venture is "proof" that the task should be shifted into the presently attainable category.

Classes of Meaning

With these two interpretations of the word "possible" in hand we may now consider two classes of meaning, each based on our two interpretations.

1. *Present Meaning.* If the determination of a degree of probability for a proposition is presently attainable, then the proposition has present meaning. This statement allows considerable latitude, which we must have in order to allow scientists to work on problems which have a low probability of being satisfactorily solved as well as on straightforward, cut-and-dried problems. If one can conduct an experiment in which a given hypothesis can be tested, and the probability of its being true can be ascertained with the tools that are presently at hand, then clearly the hypothesis has present meaning. If we cannot now conduct an experiment to test the hypothesis, it does not have present meaning.

2. *Potential Meaning.* A proposition has potential meaning if it *may* be possible to determine a degree of probability for it at some time in the future, if the degree of probability is potentially attainable. Such a proposition does not have present meaning. But as we improve our techniques and invent new ones it may become possible to test it later. Within this category we also want to allow wide latitude. There may be statements for which we know with a high degree of certainty how we will eventually test them, though we simply cannot do it now. At the other extreme are statements that we have a good deal of trouble imagining the procedures by which they will eventually be tested, but we are not ready to say that some bright soul will not some day come up with the appropriate tools.

On the basis of the above considerations, we may now formulate our principles of action with regard to hypotheses. First, since the psychologist conducting experiments must work only on problems that have a possibility of being solved with the tools that are immediately available, he must apply the criterion of present meaning in his everyday work. Therefore, only if a hypothesis has present meaning should it be subjected to experimental test. The psychologist's problems which do not have present meaning, but do have potential meaning, should be set aside in a "wait and see" category. When sufficient advances have been made so that the problem can be investigated with the tools of science, it becomes presently meaningful and can be solved. If sufficient technological advances are not made, then the problem is maintained in the category of potential meaning. On the other hand, if advances show that the problem that is set aside proves to have no potential meaning, it should be discarded as soon as this becomes evident, for no matter how much science advances, no solution will be forthcoming.

Some Examples of Meaningless Problems

With these general criteria in hand, let us investigate the problem of meaning in greater detail. First, the clearest type of meaningless proposition is the one that has no empirical reference, that does not deal with properties which can be observed by the ordinary senses of human beings. Such propositions have variously been labeled "superempirical," "mystical," "metaphysical" propositions. They are presumably formed with the use of postulated capacities such as "intuition," "divine vision," "revealed truth," and similar mystical powers. Two examples of such a proposition are "Ra (the Egyptian Sun God) has a green beard," [5] and the more hackneyed "An infinite number of angels can dance on the head of a pin." Many theological propositions fall into this category. It should be emphasized that while such propositions are labeled meaningless, science does not call them false. Such propositions just do not fall within the scope of scientific inquiry. They may be true or false (an approximation to probable or improbable as indicated earlier) by some other criterion of meaning than that which we have offered above; they are only scientifically meaningless, but may be meaningful by some "superempirical" criterion of meaning, whatever that might be.

Second, a number of propositions are used by scientists in spite of the fact that they appear to be meaningless. While our aim is to exclude *totally* meaningless propositions from science, we must not be too hasty in judging certain propositions. Some, for example, may be presently meaningless but have potential meaning. These might eventually be very fruitful. There are many categories of propositions, or concepts that we may regard as abbreviations for propositions, that are proper candidates for dismissal, though they vary in the status of their candidacy. One category that might be considered here includes some of the psychoanalytical notions as they are traditionally used; e.g., libido, id, ego, superego, etc. The view expressed by some writers is that these concepts are at least presently meaningless and by that criterion cannot be studied

---

[5] This example provides a good opportunity to say that our consideration of meaning excludes analytic statements (see Chapter 3). Since such statements are formally true, we would have to admit them as meaningful. But this consideration is really irrelevant to our present purpose for we are attempting to be practical in the sense that we are discussing the empirical testing of propositions. If one wanted to reformulate this example ("Ra by definition has a green beard") and submit it as a "theological tautology," we could not disagree. But since our purpose would be to empirically test such a statement, which we could not do, we must say that it is meaningless and that such an adversary is violating the spirit of our rules.

in their present form. According to this view, the question of how scientifically to test such concepts right now is unanswerable; and the suggestion is that they should be excluded from psychology. (We might, rather, say that they have potential meaning.) It should be hastily added that there are undoubtedly a number of psychoanalytic notions that are meaningful, and that most scientists would prefer to keep because they can be scientifically tested. The concept of repression may be one such example.

"Lower animals are conscious." This is clearly a presently meaningless proposition because there is no way to test it. It does, however, probably have potential meaning (assuming that we could agree on an adequate definition of "consciousness"). For it is entirely possible that at some later date a person will develop a "consciousness-viewing machine" that we can attach to an organism's cranium to allow us to "see" his consciousness, just as we get a visual and auditory message from a television set. Of course one would then have to determine that the display on the "machine" is actually what we have defined as the consciousness of the subject, but this matter seems relatively straightforward; e.g., coordinating the "pictures" with verbal reports. A problem similar to the above is an old one: "If you attach the optic nerve to the auditory areas of the brain will you sense visions auditorily?"

Another candidate for dismissal from psychology is a particular attempt to explain reminiscence, a phenomenon that may appear under certain conditions. To illustrate briefly, let us say that a person practices a task such as memorizing a list of words, though the learning is not perfect. He is tested immediately after a certain number of practice trials. Then, after a period of time during which he has no further practice he is tested on the list of words again. On this second test it may be found that he recalls more of the words than he did on the first test. This is *reminiscence*. We may say that reminiscence occurs when the recall of an incompletely learned task is more complete after a period of time has elapsed following the learning period than it is immediately after the learning period. The problem is how to explain this phenomenon.

One possible explanation of reminiscence is that, while there are no formal practice trials following the initial learning period, the subject continues implicitly to practice the task. That is, he "rehearses" the task "to himself," following the initial practice period, and before the second test. This informal rehearsal could well lead to a higher score on the second test. Now our purpose is not to take issue with this attempt to explain reminiscence. Rather, we wish to examine a line of reasoning leading to a rejection of "rehearsal" as an explanation of the phenomenon.

One psychologist lists several types of evidence which suggest that rehearsal cannot account for reminiscence. Among this evidence is the following statement: "Rats show reminiscence in maze learning (Bunch and Magsdick, 1933), and it is not easy to imagine rats rehearsing their paths through a maze between trials" (Deese, 1952, p. 175). With our "consciousness-viewing machine" we could test this statement. Without such a device, the present author's inclination is to apply the criterion of present meaning to this statement. Such a statement cannot seriously be considered as bearing on the problem of reminiscence—there is simply no way at present to determine whether rats do or do not rehearse, assuming the common definition of "rehearse." (This does not mean, of course, that other explanations of reminiscence are meaningless.)

Take a situation in which a clinical patient cannot talk. He scores low on an intelligence test. After considerable clinical work, he is given another intelligence test and registers a significantly higher score. Did the intelligence of the patient increase as a result of the clinical work? Or was the first administration faulty because of the difficulties of administering the test to a non-verbal patient? These presently meaningless questions may have potential meaning.

Let us consider testing two theories of forgetting: the *disuse theory,* which says that forgetting occurs strictly because of the passage of time, and the *interference theory,* which says that forgetting is the result of competition from other learned material. Which theory is more probably true? An experiment by Jenkins and Dallenbach (1924) is frequently cited as evidence in favor of the interference theory, and this is scientifically acceptable evidence. This experiment showed that there is less forgetting during sleep (when there is little interference) than during waking hours. However, their data indicate considerable forgetting during sleep, which is usually accounted for by saying that even during sleep there is some interference (possibly from incoming stimuli, dreaming, etc.). To determine whether or not this is so, to test the theory of disuse strictly, we must have a condition in which a person has zero interference. Technically, there would seem to be only one condition which might satisfy this requirement—death—and even this is doubtful for it would be difficult to measure retention under such a condition. Thus, the Jenkins-Dallenbach experiment does not provide a completely adequate test of the theory of disuse. Therefore, the proposition that "during a condition of zero interference there will be (or will not be) forgetting" is a presently meaningless proposition, although it has potential meaning.

The interested reader should consider for himself a number of other questions in psychology to determine whether they are meaningful. For example: "A subject in an experiment performs exactly as he would if he were not in an experiment." Is it presently meaningful to ask how the subject would perform if the apparatus or the questionnaire or the test were not used?

There is still another type of problem which we must classify as meaningless only because it has been inadequately formulated. For instance, the college instructor in psychology must face the fact that students will continue to ask such questions as "What's the matter with his (my, your) mind?" "I came into this course to find out how the mind works," etc. A number of such questions are meaningless because there is no way to obtain relevant observations. But sometimes such questions can be reformulated in such a way that they become meaningful. This involves defining the concepts contained in the questions, or substituting definable terms for those which are not. For example, we could reformulate the question "What's the matter with my mind," probably after a lengthy discussion with the student to determine just what is meant, so that it might take the following form: "Why am I compelled to count the number of door knobs in every room that I enter?" Such a question is still difficult to answer, but at least the chances for an answer are better in that the question is more precisely stated and refers to events that are more readily observable. This issue of formulating adequate problems is closely related to our later discussion of operationally defining concepts (Chapter 4) and of formulating hypotheses (Chapter 3).

Some Additional Considerations of Problems

Even after we have determined that a problem is presently meaningful, there are other requirements to be met before considerable effort is expended in conducting an experiment. One such requirement is that the problem be sufficiently important. Numerous problems arise for which the psychologist can furnish no answers immediately or even in the future, though they are in fact meaningful problems. Some problems are just not important enough to justify research—they are either too trivial, or too expensive (in terms of time, effort, money, etc.), to answer. For instance, the problem of whether rats prefer Swiss or American cheese is likely to go unanswered for centuries; similarly, "why men fight" not because it is unimportant, but because its answer would require much more effort than society is willing to expend on it.

Sometimes the psychologist is aware of a problem that is meaningful, adequately phrased, and important, but an accumulation of experiments on the problem show contradictory results. If we could rule out interactions (see p. 211) as accounting for the contradictory results, there would seem to be no reason for such discrepancies. This is what might be called "the impasse problem." When faced with this situation, it would not seem worthwhile to conduct "just another experiment" on the problem, for little is likely to be gained, regardless of how the experiment turns out. That is, if the experiments are numerous and contradictory, little is to be gained by just chalking up one more set of data on either side of the fence. Unless the experimenter can be extremely imaginative and develop a totally new approach which has some chance of systematizing the knowledge in the area, it is likely he should stay out of it and use his energy to perform research on a problem that has a greater chance of contributing some new knowledge.

Some aspects of this general discussion may strike you as representing a "dangerous" point of view. One might ask, how we can ever know that a particular problem is really unimportant. Perhaps the results of an experiment on what some regard as an unimportant problem might turn out to be very important—if not today, perhaps in the future. Unfortunately, there is no answer to such a position. Such a situation is, indeed, conceivable, and our position as stated above would "choke off" some important research. Other things being equal, however, the author would suggest that if an experimenter can foresee that his experiment will have some significance for theory or for some applied practice, his results are going to be more valuable than if he cannot foresee such consequences. There are some psychologists who would never run an experiment unless it has some specific influence on a given theoretical position. This might be too rigid a position, but it does have merit.

There is no clear delineation between an important problem and an unimportant one, but it can be fairly clearly established that some problems are more likely to contribute to the advancement of psychology than are others. And it is a good idea for the experimenter to try to choose what he considers an important problem rather than a relatively unimportant problem. Within these rather general limits, no further restrictions are suggested. In any event, science is the epitome of the democratic process, and any scientist is free to work on whatever problem he chooses. A lot of experiments assuredly might as well go right down the drain, but what some scientists would judge to be "ridiculous problems" may well turn out to have revolutionary significance. Some psychologists have expressed a wish for there to be created a professional journal with a title like "The Journal of Crazy Ideas," to encourage wild and speculative research.

3

The Hypothesis

The Nature of a Hypothesis

We have previously noted that a scientific investigation starts with the statement of a meaningful problem. Following this, a tentative solution to that problem is offered in the form of a proposition. The proposition must be meaningful—it must be possible to determine whether it is probably true or false. Thus, a hypothesis is a meaningful proposition that *may be* the solution of a problem. To enlarge on this definition of a hypothesis, briefly consider the relationship between the problem and the hypothesis.

First, if it is found after suitable experimentation that the relevant hypothesis is probably true, then we may say that the hypothesis *solves* the problem to which it is addressed. If the relevant hypothesis is probably false, we may say that it does not solve the problem. To illustrate, let us consider the problem: "Who makes a good bridge player?" Our hypothesis might be that "people who are intelligent and who show a strong interest in bridge make good bridge players." The collection and interpretation of sufficient data might confirm the hypothesis. In this case we can say that we have solved the problem because we can answer the question. (Of course, we might add that the problem is not *completely* solved. We are using "solved" in an approximate sense and further research is required to enlarge our solution, as, for instance, finding other factors that make good bridge players. In this case we may arrive at a hypothesis that has a greater probability than the earlier one; and it offers a more complete solution.) On the other hand, let us say that we fail to confirm our hypothesis. In this event it should be apparent that we have not solved the problem, i.e., we have not obtained any information on what qualities make a good bridge player.

Frequently when we obtain a true hypothesis and thus solve a problem we may say that the hypothesis *explains* the phenomena with which the problem is concerned. Let us assume that a problem exists for the reason that we are in possession of a certain fact. As noted in the last chapter, this fact, existing in isolation, requires an explanation. The need to explain the fact is our problem. If we can relate that fact to some other fact in an appropriate manner, we can say that the first fact is explained. A hypothesis is the tool by which we seek to accomplish such an explanation. That is, we use a hypothesis to state a possible relationship between one fact and another. If we find that the two facts are actually related in the manner stated by the hypothesis, then we have accomplished our immediate purpose—we have explained the first fact (a more precise discussion of explanation is offered in Chapter 13).

To illustrate, let us return to an example used in Chapter 2. Becquerel was presented with the fact that a certain photographic film had been fogged. This fact demanded an explanation. In addition to noting this fact, Becquerel also noted a second fact: that a piece of uranium was lying near the photographic film. His hypothesis was that some characteristic of the uranium produced the fogging. Becquerel's test of this hypothesis proved successful—he established that his hypothesis was true. Thus, by relating the fogging of the film to a characteristic of the uranium he explained his fact.

But what do we mean by "fact." "Fact" is a common-sense word, and as such its meaning is rather vague. We understand something by it, such as "a fact is an event of actual occurrence"—it is something that we are quite sure has happened (Becquerel was quite sure that the photographic film was fogged). But it may facilitate our task if we replace this common-sense word with a more precise term. For instance, instead of using the word *fact*, suppose that we conceive of the fogging of the film as a *variable*, that is, the film may be fogged in varying degrees, from a zero amount up to some large amount, such as total exposure. Similarly, the amount of radioactive energy that is given off by a piece of uranium may be conceived of as a variable—it may be an amount anywhere from zero to a large amount. Therefore, instead of saying that two *facts* are related, we may now make the more desirable statement that two *variables* are related. The advantages of this procedure are sizeable. For example, we may now hypothesize a *quantitative* relationship—the greater the amount of radioactive energy given off by the piece of uranium, the greater the fogging of the photographic plate. Hence, instead of making the rather crude distinction between "fogged" and "unfogged" film, we may now talk about the *amount* of fogging. Similarly, the uranium is not simply giving off radioactive energy, it is emitting an *amount* of energy. We are now in a position to make statements of wide applicability. Before, we could only say that if the uranium gave off energy, the film would be fogged. Now, we can say, for instance, that if the uranium gives off a "little" energy, the film will be fogged a small amount; if the uranium gives off a lot of energy, the film will be greatly fogged, and so on. Of course, this example is only illustrative—we can make many more statements about the relationship of these two variables with the use of numbers. But this matter will be taken up more thoroughly shortly.

These considerations now allow us to enlarge on our preceding definition of a hypothesis. For now we may define a hypothesis as *a meaningful statement of a potential relationship between two (or more) variables.*

There are a number of terms used in science other than "hypothesis" to refer to statements of relationships between variables. They are such words as "theories," "laws," etc. The discussion at the present level will be applicable to any statement of a relationship between variables—hypotheses, theories, laws, or whatever. Sometimes we will use the terms "principles" or "empirical generalizations," in which case we mean a hypothesis *or* a "theory" *or* a "law." We might add one final qualification: the variables under consideration are *empirical*—they refer to observable aspects of nature. Thus, when we use the word "hypothesis," we understand that it is an empirical hypothesis—an hypothesis that refers to data that we can obtain from our observation of nature.

Synthetic, Analytic, and Contradictory Statements

This last point above is important for the understanding of the nature of a hypothesis. It will be advantageous to consider it more fully. To accomplish this let us note that all possible statements fall in one of three categories: *synthetic, analytic,* or *contradictory.* These three kinds of statements differ on the basis of their possible *truth values.* By *truth value* we mean whether a statement is true or false. Thus, we may say that a given statement has the truth value of *true* (such a statement is "true") or that it has the truth value of *false* (this one is "false"). Because of the nature of their construction (the way in which they are formed), however, some statements can take on only certain truth values. Some statements, for instance, can take on the truth value of *true* only. Such statements are called analytic statements (other names for them are "logically true statements" or "tautologies"). Thus, an analytic statement is a statement that is always true—it cannot be false. The statement "Either you are older than your brother or you are not older than your brother" is an example of an analytic statement. Such a statement exhausts the possibilities, and since one of the possibilities must be true, the statement itself must be true. A contradictory statement (sometimes also called a "self contradiction" or a "logically false statement"), on the other hand, is one that always assumes a truth value of false. That is, because of the way in which it is constructed, it is necessary that the statement be false. A negation of an analytic statement is obviously a contradictory statement. For example, the statement "It is false that you are older than your brother or you are not older than your brother (or the logically equivalent statement "You are older than your brother and you are not older than your brother") is a contradictory statement. Such a statement includes all of the logical possibilities, but says that all of these logical possibilities are false.

The third type of statement is the synthetic statement. A synthetic statement may be defined as any statement that is neither an analytic nor a contradictory statement. In other words, a synthetic statement is one that may be *either* true or false, or more precisely, it has a probability of being true or false. An example of a synthetic statement would be "You are older than your brother," a statement that may be either true or false. And the important point to observe in this discussion is that a *hypothesis must be a synthetic statement*. Thus any hypothesis must be capable of being proven (probably) true or false.

The differences among these three types of statements are highlighted in Table 1.1. There we may see that the symbolic statement of an ana-

Table 1.1

Possible Kinds of Statements*

Type of Statement	Analytic	Synthetic	Contradictory
Symbolic Statement	*a* or not *a*	*a*	*a* and not *a*
Example of Statement	I am in Chicago, or I am not in Chicago.	I am in Chicago.	I am in Chicago and I am not in Chicago.
Truth or Probability Value	(Absolutely) true	1.0 > P > 0	(Absolutely) false

* The symbol > may be read as "greater than" or < as "less than." Thus, for the synthetic statement, P is less than 1.0 but zero is less than P. Or alternatively, 1.0 is greater than P; but P is greater than 0.

lytic proposition is "*a* or not *a*." For instance, if we let *a* stand for the sentence "I am in Chicago," then the appropriate analytic statement is "I am in Chicago or I am not in Chicago." This statement is necessarily true because no other possibilities exist. The synthetic statement is symbolized by *a*. Thus, following our example, it says "I am in Chicago," and the probability of this statement is necessarily less than (<) 1.0, but necessarily greater than (>) 0. The symbolic statement of the contradictory type of proposition is "a and not a"—"I am in Chicago and I am not in Chicago." Clearly, such a statement is absolutely false, barring such unhappy possibilities as my being in a severed condition.

Now, why should we state hypotheses in the form of synthetic statements? Why not use analytic statements, in which case it would be guaranteed that our hypotheses are true? The answer to such a question appears in an understanding of the function of the various kinds of statements. The reason that a synthetic statement may be true or false is that it refers to the empirical world, i.e., it is an attempt to tell us something about nature. And as we previously saw, every statement that refers to natural events might be in error. An analytic statement, however, is empty. That is, while it is absolutely true, it tells us nothing about the empirical world. This characteristic results because an analytic statement includes all of the logical possibilities, but it does not attempt to inform us which is the true one. And this is the price that one must pay for absolute truth. If one wishes to state information about nature he must use a synthetic statement, in which case the statement always runs the risk of being false. Thus, if someone asks me if you are older than your brother I might give him my best judgment, say, "you are older than your brother" which is a synthetic statement. I may be wrong in this statement, but at least I am trying to tell him something about the empirical world. Such is the case with our scientific hypotheses—they may be false in spite of our efforts to assert true hypotheses, but they are cognitive (contain information or content) in the sense that they say something about nature.

Now, if analytic statements are empty and thus do not tell us anything about nature, why do we bother with them in the first place? The answer to this question could be made quite detailed. Suffice it to say here that analytic statements are quite valuable in facilitating deductive reasoning (inferences). The statements in mathematics and logic are analytic and contradictory statements, and are valuable because they allow us to transform synthetic statements without adding additional knowledge (recall that analytic statements are empirically empty). The essential point is that sciences use all three types of statements, but use them in different ways. We have said that the synthetic type of proposition is used for the statement of hypotheses; for, in stating a hypothesis, we attempt to say something informative about the natural world. This attempt carries with it the possibilities that our hypothesis is probably true or false.

The Manner of Stating Hypotheses

Granting, then, that a hypothesis is a statement of a potential empirical relationship between two or more variables, and also that it is possible to determine whether the hypothesis is probably true or false, we might well ask what form that statement should take. That is, precisely how should we state hypotheses in scientific work?

Lord Russell answers this question by proposing that the logical form of the general implication be used for expressing hypotheses (cf. Reichenbach, 1947, p. 356). Using the English language, the general implication may be expressed as: "*If* . . . , *then* . . ." That is to say, "*if* certain conditions hold, *then* certain other conditions should also hold." To better understand the "If . . . then . . ." relationship, let *a* stand for the first set of conditions, and *b* for the second set of conditions. In this case the general implication would be "If *a*, then *b*." But, in order to communicate what the conditions indicated by *a* are, we must make a statement. Therefore, we shall consider that the symbols *a* and *b* are actually statements which express these two sets of conditions. And if we join these two simple statements, as we do when we use the general implication, then we end up with a single compound statement. This compound statement *is* our hypothesis.

The statement *a*, incidentally, is referred to as the *antecedent condition* of the hypothesis (it "comes first"), while *b* is called the *consequent condition* of the hypothesis (it follows the antecedent condition). We have previously noted that a hypothesis is a statement relating two variables. Since we have said that antecedent and consequent conditions of a hypothesis are stated as propositions, it follows that the symbols *a* and *b* are *propositional variables*. A hypothesis, thus, proposes a relationship between two (propositional) variables by means of the general implication as follows: "If *a* is true, then *b* is true." The variables *a* and *b* may stand for whatever we wish. If we suspect that two particular variables are related, we might hypothesize a relationship between them. For example, we might think that industrial work groups that are in great inner conflict have decreased production levels. Here the two variables are (1) the amount of inner conflict in an industrial work group and (2) the amount of production that work groups turn out. We can formulate two sentences: (1) "An industrial work group is in great inner conflict"; and (2) "That work group will have a decreased production level." If we let *a* stand for the first statement, and *b* for the second, our hypothesis would read: "*If* an industrial work group is in great inner conflict, *then*, that work group will have a decreased production level."

With this understanding of the general implication for stating hypotheses, it is well to inquire about the frequency with which Russell's suggestion has been accepted in psychology. The answer is quite clear: the explicit use of the general implication is almost nonexistent. Two samples of hypotheses essentially as they are stated in professional journals should suffice to illustrate the point:[1]

1. The present investigation is designed to study the effects of the teacher's praise on reading growth. (Silverman, 1957)

2. Giving students an opportunity to write comments on objective

---

[1] The statement of these hypotheses has been modified somewhat for easier comprehension. The interested reader, of course, may consult the original articles.

examinations results in higher test scores (McKeachie, Pollie & Speisman, 1955).

Clearly these hypotheses, or implied hypotheses, fail to conform to the form specified by the general implication. Is this bad? Are we committing serious errors by not precisely heeding Russell's advice? Not really. For as Hempel and Oppenheim (1948) point out, it is always possible to restate such hypotheses as general implications. For example, the above hypotheses could be restated as follows.

The first hypothesis contains two variables: (1) amount of praise and (2) amount of reading growth. The propositions concerned with these variables are (1) A teacher praises a student for good reading performance, and (2) the student's reading growth increases. The hypothesis relating these two variables may now be expressed. *"If* a teacher praises a student for good reading performance, *then* the student's reading growth will increase."

We may similarly identify the variables contained in the second hypothesis and state the propositions as follows: (1) Students are given the opportunity to write comments on objective examination questions, and (2) those students achieve higher test scores. The hypothesis: *"If* students are given the opportunity to write comments on objective examination questions, *then* those students achieve higher test scores."

It is apparent that these two hypotheses fit the "If *a* then *b*" form, although it was necessary to modify somewhat the original statements. Even so, these modifications did not change the nature or the meaning of the hypotheses.

What we have said to this point, then, is that Russell has suggested the use of the general implication for stating hypotheses, and that psychologists do not take his advice in that they express their hypotheses in a variety of ways. However, we can restate their hypotheses as general implications. The next question, logically, is why did Russell offer this advice, and why are we making a point of it here? Briefly, the way in which we determine whether or not a hypothesis is confirmed depends on our making certain inferences from experimental findings to the hypothesis. The rules of logic tell us what kind of inferences we can legitimately make (they are called valid inferences). But in order to determine whether or not the inferences are valid the statements involved in the inferences (e.g., the hypotheses) must be stated in certain standard forms, one of which is the *general implication*. Hence, in order to discuss the nature of experimental inferences and really to understand them, we must use standard logical forms. This area will be covered more completely in Chapter 11, when we consider the nature of experimental inferences.

There is yet another form that is used in the stating of hypotheses. It involves certain mathematical statements that are essentially of the following nature: X = f(Y). That is, a hypothesis stated in this way proposes that some variable (X) is related to some variable (Y), or, alternatively, that X is a function of Y. Such a mathematically stated hypothesis clearly fits our more general definition of a hypothesis to the effect that two variables are related. While the variables in this case are quantitive (their values can be measured with numbers), the two variables may still refer to whatever we wish. For instance, we might refer to hypothesis 2 on p. 33 and assign numbers to the independent variable. The independent variable, which would be Y, might be stated as the extent to which students are given the opportunity to write comments on examination questions. For instance, we might develop a scale such that 1 would indicate very little opportunity, 2 a little greater opportunity, 3 a medium amount of opportunity, and so on. Test scores, the dependent variable (X), could be similarly quantified—100 might be the highest possible score and zero the lowest. Thus, the hypothesis could be tested for all possible numerical values of the independent and the dependent variables.

In any event, the important point here is that even though a hypothesis is stated in a mathematical form, that form is basically of the "If a, then b" relation. Instead of saying "If a, then b" we merely say "If (and only if) Y is this value, then X is that value." For example, if X is 3 (a medium opportunity to write comments) then Y is 75 (an average grade).

Consider two common misconceptions about the statement of hypotheses as general implications. FIRST, it may be said that the antecedent conditions *cause* the consequent conditions. This may or may not be the case. The general implication merely states a potential relationship between two variables—*if* one set of conditions holds, *then* another set will be found to be the case—not that the first set causes the second. Thus, if the hypothesis is in fact highly probable, we can expect to find repeated occurrences of both sets of conditions together.

SECOND, the general implication does not assert that the consequent conditions are true. Rather, it says that *if* the antecedent conditions are true, *then* the consequent conditions are true. For example, the statement, "*If* I go downtown today, *then* I will be robbed" does not mean that I *will be* robbed. Even if the compound statement is true, I might not go downtown today. Thus, *if* the hypothesis is highly probable then whether or not I will be robbed depends on whether or not I satisfy the antecedent conditions.

Types of Hypotheses

We have suggested that the general implication is a good form for stating hypotheses. We have discussed the use of the implication, but have said nothing explicitly about the generality of the implication. In one of the previously cited examples it was said that if an industrial work group has a certain characteristic, certain consequences follow. We have not specified what industrial work group, but we have said that the hypothesis concerns at least *some* industrial work groups out of all possible groups. Are we now justified in asserting that the hypothesis holds for all industrial work groups? The answer to this question is ambiguous, and there are two possible courses. First, we could say that the particular work group out of all possible work groups with which we are concerned is unspecified, thus leaving the matter up in the air. Or, second, we could assume that we are asserting a universal hypothesis, i.e., that it is implicitly understood that we are talking about *all* industrial work groups that are in conflict. In this instance, if you take *any* industrial group in conflict, the consequences specified by the hypothesis should follow. In the interest of the advancement of knowledge, we lean toward the latter interpretation, for if the former interpretation is followed, no definite commitment is made on the part of the scientist, and if nothing is risked, nothing is gained. If it is found that the hypothesis is not universal in scope (that it does not apply to *all* industrial work groups), it must be further limited. This is at least a definite step forward. That this is not an idle question is made apparent by many recent psychological articles. Hull, for example, states a number of his empirical generalizations in this manner. His Postulate IV says: "If reinforcements follow each other at evenly distributed intervals . . . the resulting habit will increase in strength . . ." (Hull, 1952, p. 6). Without worrying here about what the specific variables in Postulate IV mean, we may observe the general form of the principle. Is it clear that Hull is asserting some relationship between *all* reinforcements and *all* habits? It is by no means, but the most efficient course open to us, as we have discussed it above, would be to *assume* that he is asserting such a universal relationship.

While it is recognized that the goal of the scientist is to assert his hypotheses in as universal a fashion as possible, it is also clear that he should explicitly state the degree of generality with which he is asserting his hypothesis. With this in mind, let us investigate the possible types of hypotheses which the scientist has at his disposal.

The first type of hypothesis is what is called the *universal hypothesis*, which asserts that the relationship in question holds for all variables, for all time, and at all places. An example of a universal hypothesis would be "For all rats, if they are rewarded for turning left, then they will turn left in a T maze."

Another type of hypothesis is the *existential hypothesis,* which asserts that the relationship stated in the hypothesis holds for at least one particular case ("existential" implies that one exists). For instance, "There is at least one rat, that if he is rewarded for turning left, then he will turn left in a T maze."[2]

Now, why would a psychologist ever want to use the existential form for stating hypotheses? The author would hesitate to attempt to explore here the possible reasons a scientist might have for doing this, but he would have no reluctance whatsoever to challenge you to search the literature for examples of this type of hypothesis. One would not have to go very far, for instance, to note Hull's Postulate V (D) which says "At least some drive conditions tend partially to motivate into action habits which have been set up on the basis of different drive conditions" (Hull, 1952, p. 7). Here again, we need not be concerned with what the postulate actually says *empirically;* we should merely note that it is in the form of the existential hypothesis. (It might be better to say "At least one drive condition . . ." to make it clear that Hull's hypothesis assumes the existential form as previously discussed.) It may be added that this form of hypothesis can be very useful in psychological work, for many times a psychologist asserts that a given phenomenon exists, regardless of how frequently it exists. In this connection Bugelski says "often one subject is as useful as many, if the problem involved is of the 'is-it-possible' nature. Herman Ebbinghaus used himself as subject and contributed greatly to our knowledge of learning. Raymond Dodge studied his own knee jerk for several years and made notable contributions to our information about reflex action. It takes only one positive case to prove something can happen" (Bugelski, 1951, pp. 115-116). This is akin to the problem of whether purple elephants exist. No one would care to assert that all elephants are purple; however, the appropriate hypothesis would be the existential type if it is proven that at least one exists. The scientist would then have as his goal the further delimiting of conditions until he could eventually assert a universal hypothesis with a number of qualifying conditions, e.g., all elephants in a certain location who are 106 years old and who answer to the name "Tony" are purple. One unfortunate characteristic of at least some statements that are so limited is that they thereby lack predictive power. In this example, for instance, it is unlikely that an elephant that showed

---

[2] There are other types of hypotheses that could be discussed but because they are relatively rarely used they will not be considered here (cf. Hempel, 1945; Reichenbach, 1949).

up in this location at some future time would have the characteristics specified.

Arriving at a Hypothesis

It is difficult to specify the process of arriving at a hypothesis with any finality. While considerable research has been related to this problem, it is not possible at this time, or in the foreseeable future, to specify adequately just what phases a scientist goes through in arriving at a hypothesis. We *can* say that the process of arriving at a hypothesis is a creative matter which has been the object of studies in thinking, imagination, concept formation, and the like. This leads to a distinction advanced by Reichenbach, (1938) who says that the manner in which the scientist actually arrives at his hypothesis falls within the *context of discovery*, while the presentation of the proof that a hypothesis is probably true is the *context of justification*. Thus, science in general is not interested in the context of discovery (however, psychology *is* interested, since this is a portion of its subject matter). Science is concerned with the context of justification, for here, instead of presenting the thought processes as they occurred in the development of the hypothesis, the scientist reconstructs his thinking logically—he sets forth a justifiable set of inferences that lead from one statement to another. The adequate expression of our thoughts in science is through the rational reconstruction of those thoughts. When the scientist publishes his hypothesis, and material related to it, he does not relate how he actually arrived at the hypothesis—he does not say that "while I was sitting in the bathtub the following hypothesis occurred to me . . ." Rather he justifies his hypothesis. What he writes falls within the context of justification and of course is discussed in the major part of this book. In this connection we might note that some critics of the study of scientific methodology (or philosophy of science) say that such an endeavor is worthless. They say that scientific discoveries are not made in a logical, step-by-step fashion in strict conformity with the scientific method; a scientist does not sit down with a problem and rationally go through the phases of the scientific method as listed in Chapter 1. To this criticism we must answer that this may be true, or in fact it may not be true, but whether it is true or false is really irrelevant. What is relevant is that when the scientist communicates his findings to others, he utilizes the context of justification. Whether or not the use of the context of justification will facilitate the discovery of hypotheses is an empirical question for studies of thinking. It is possible, however, that the making of valuable discoveries can be facilitated by studying the scientific method, for it seems reasonable that if we learn how scientific discoveries have been made in the past, and set such procedures down in a systematic way, we may be able to make new discoveries more efficiently in the future.

Dealing further with the context of discovery, we may note that when a scientist arrives at a hypothesis he surveys a mass of data (implicitly or explicitly), abstracts certain aspects of it, sees some similarities in the abstractions, and relates the similarities in order to arrive at generalizations. For instance, the psychologist particularly observes stimulus and response events. He notes that some stimuli are similar to other stimuli and that some responses are similar to other responses. He defines as belonging to the same class those stimuli that he has noticed as being similar according to a certain characteristic, and similarly for the responses. Consider the following situation in a Skinner Box. A rat presses a lever and receives a pellet of food. At about the time the rat presses the lever a click is sounded. After a number of associations between the click, pressing the lever, and eating the pellet, the rat will learn to press the lever when a click is sounded.

In this situation the experimenter might judge that all of the separate instances of the lever-pressing response are sufficiently similar for them to be classified together. He thus forms a class of lever-pressing responses out of a number of similar lever-pressing response instances. In like manner he might form a class of all of the stimulus instances of clicks, judging that all of the clicks are similar enough to form a general class. It can thus be seen that the psychologist uses classification to distribute a large amount of data into a smaller number of categories that can be handled efficiently. He facilitates the handling of these data by assigning symbols to the classes. He then attempts to formulate relationships between the classes. By "guessing" that a certain relationship exists between the classes, he formulates his hypothesis. For example, he might suspect that when a click is made, a certain response will follow. His hypothesis might be like the following: If a click stimulus is presented a number of times to a rat in a Skinner Box, and if the click is frequently associated with pressing a lever and eating a pellet, then the rat will press the lever in response to the click on future occasions. Parenthetically, this seems to be typical of the process that the scientist goes through, more or less as an ideal. Some scientists, probably the more compulsive ones, go through each step in considerable detail, while others do so in a more haphazard fashion. But whether or not scientists go through these steps in arriving at a hypothesis, explicitly, or implicitly, they all seem to approximate them to some extent.

We may note the views of some others on this topic. Homer Dubs writes: "It is a well known fact that most hypotheses are derived from analogy . . . Indeed, careful investigations will very likely show that all philosophic theories are developed analogies" (Dubs, 1930, p. 131). In support he points out that Locke's conception of simple and complex ideas was probably suggested by the theory of chemical atoms and compounds that was becoming prominent in his day. Underwood has written: "How does one learn to theorize? It is a good guess that we learn this skill in the same manner we learn anything else—by practice" (Underwood, 1949, p. 17).

Of course, some hypotheses are more difficult to formulate than others. It seems reasonable to say that the more general a hypothesis is, the more difficult it is to conceive. The more general hypotheses must await the genius to proclaim them, at which time a science usually makes a sizable spurt forward, as happened in the cases of, say, Newton and Einstein. It appears that, to formulate useful and valuable hypotheses, a scientist needs, first, sufficient experience in the area and, second, the quality of "genius." The main problem in formulating hypotheses in complex and disorderly areas is the difficulty of establishing a new "set"—the ability to create a new solution that runs counter to, or on a different plane from, the existing knowledge. This is where scientific "genius" is required.

Consider the source of hypotheses from a somewhat different position, in particular with reference to the results of a scientific inquiry. The findings can be regarded, in a sense, as a stimulus to formulate new hypotheses—while results may be used to test a hypothesis, they can also suggest new hypotheses. For example, if the results indicate that the hypothesis is false, they can be used to form a new hypothesis that is in accord with the results of the experiment. In this case the new hypothesis must be tested in a new experiment. Now, what happens to a hypothesis if it turns out to be false? If there is a new (potentially better) hypothesis to take its place, it can be readily discarded. But if there is no new hypothesis, then we are likely to maintain the false hypothesis, at least temporarily, for no hypothesis ever seems to be finally discarded in science unless it is replaced by a new one.

Criteria of Hypotheses

After we have formulated our hypothesis (or better, our hypotheses), we must determine whether or not the hypothesis is a "good" one. Of course, we eventually will test our hypothesis to determine whether the data confirm or disconfirm it, and certainly, other things being equal, a confirmed hypothesis is better than a disconfirmed hypothesis in the sense that it offers one solution to a problem and thus provides some additional knowledge about nature. But even so, some confirmed hypotheses are better than other confirmed hypotheses. We must now inquire into what we here mean by "good" and by "better." To answer this question we offer the following criteria by which to judge hypotheses. Each criterion should be read with the understanding that the hypothesis that meets it to the greatest extent is the best hypothesis, assuming that the hypothesis satisfies the other criteria equally well and that the criteria are about equal in worth. It should also be understood that these are flexible criteria. They are offered somewhat tentatively, and as the information in this important area increases, they will no doubt be modified. The hypothesis:

1. . . . must be meaningful. The hypothesis that has present meaning is superior to the hypothesis that his only potential meaning.

2. . . . should be in general harmony with other hypotheses in the field of investigation. While this is not essential, in general the disharmonious hypothesis has the lower degree of probability. For example, a medical doctor recently advanced the hypothesis that eye color is related to certain personality characteristics. This hypothesis is at an immediate disadvantage because it conflicts with the existing body of knowledge. We know, for instance, that hair color has never been demonstrated to be related to personality traits. There is considerable additional knowledge of this sort which, taken together, would suggest that the "eye color" hypothesis is not true—it is not in harmony with what we already know.

3. . . . should be parsimonious. If two hypotheses are advanced to answer a given problem, the more parsimonious one is to be preferred. For example, if we have evidence that a person has correctly guessed the symbol (hearts, clubs, etc.) on a number of cards significantly more often than chance, we might advance several hypotheses to account for this phenomenon. One might be to postulate extrasensory perception (ESP) and another to say that the subject "peeked" in some manner. Clearly the latter would be more parsimonious. This principle or ones similar to it, has been previously expressed in various forms. For instance, William of Occam advanced a rule (called *Occam's razor*) to the effect that entities should not be multiplied without necessity. A similar rule was expressed by G. W. Leibniz' principle of the identity of indiscernibles. Lloyd Morgan's canon is an application of the principle of parsimony to psychology: "In no case is an animal activity to be interpreted in terms of higher psychological processes, if it can be fairly interpreted in terms of processes which stand lower in the scale of psychological evolution and development" (Morgan, 1906, p. 59). It is apparent that both principles have the same general purpose—they both seek, other things being equal, the most parsimonious explanation of a problem (cf. Newbury, 1954). Thus, we should not prefer a complex hypothesis if a simple one has equal explanatory power; we should not use a complex concept in a hypothesis (e.g. ESP) if a simpler one will serve as well (e.g., peeking

at the cards); we should not ascribe higher capacities to organisms if the postulation of lower ones can equally well account for the behavior to be explained.

4. . . . should answer (be relevant to) the problem, if it is true. It would seem unnecessary to state this criterion, except that examples can be found in the history of science where the right answer was given to the wrong problem. It is often important to make the obvious explicit.

5. . . . should have logical simplicity. By this we mean logical unity and comprehensiveness, not ease of comprehension. Thus, if one hypothesis can account for a problem by itself, and another hypothesis can also account for the problem but requires a number of supporting hypotheses or *ad hoc* assumptions, the former is to be preferred because of its greater logical simplicity (cf., Cohen and Nagel, 1934, pp. 212-215). (The close relationship of this criterion to that of parsimony should be noted.)

6. . . . should be expressed in a quantified form, or be susceptible to convenient quantification. The hypothesis that is more highly quantified is to be preferred. The contrast between a quantified and a nonquantified hypothesis was illustrated earlier in the example from the work of Becquerel.

7. . . . should have a large number of consequences, should be general in scope. The hypothesis that yields a large number of deductions (consequences) will explain more facts that are already established, and will make more predictions about events that are as yet unstudied or unestablished (some of which may be unexpected and novel). In general it may be said that the hypothesis that leads to the larger number of deductions will be the more fruitful hypothesis.

The Guidance Function of Hypotheses

We have already discussed the ways in which hypotheses allow us to establish "truth." It is well here to ask how an inquiry gets its direction. In nature, for instance, how do we know where to start our search for "truth"? The answer is that hypotheses direct us. An inquiry cannot proceed until there is a suggested solution to a problem in the form of some kind of hypothesis.[3]

[3] Whether or not one chooses to assume that a hypothesis guides an inquiry in a strictly exploratory type of experiment (see p. 48) is an arbitrary decision. The author is taking the position here that some guidance is offered in this type of situation, that there is some reason that data of a particular kind are gathered, and whether or not there is an explicit hypothesis, the author is assuming that there is at least an implicit one, no matter how vague it might be. As Underwood says ". . . probably no one undertakes an investigation without some thought as to 'what will happen.' Such thoughts may not be verbalized, but that they are almost universally there no one can doubt" (Underwood, 1957, p. 208). And again, "Research in a relatively new area of investigation is seldom undertaken without some conceptual

Francis Bacon proposed that the task of the scientist is to classify the entire universe. But the number of data in the universe is, if not infinite, at least indefinitely large. To make observations in such a complex world, we must have some kind of guide. If not we would have little reason for not sitting down where we are and describe a handful of pebbles, or whatever else happens to be near us. We must set some priority on the kind of data that we study, and this is accomplished by the hypothesis. Hypotheses, then, serve to guide us to make observations that are pertinent to our problem—they tell us what observations are to be made, and what observations are to be omitted. If, for instance, we are interested in the problem of why a person taps every third telephone pole he passes, our hypothesis would probably guide us in the direction of a better understanding of compulsions—it would take us a long time if we started out in a random direction to solve our problem and commence, for instance, counting the number of blades of grass in a field. That this point is not universally accepted, however, is apparent when one notes that certain people believe that hypotheses are not valuable in some kinds of scientific work. They ask, for instance, what there is to guide us in selecting our hypotheses. To this we must answer that we know very little, but that we must be confident that more complete answers are, at least eventually, forthcoming as our research on the thinking process accumulates. Such a question is one that clearly lies within the context of discovery, and to place it within the context of justification is a rather serious error.

---

scheme in mind . . . without some preconception as to the nature of the phenomena and perhaps the processes lying behind them. These predilections are usually lightly held but they do afford the initial working hypotheses . . ." (Underwood, 1957, p. 258).

4

The Experimental Plan

We have noted that a scientific inquiry starts with a problem. The problem must be *meaningful* and may be stated in the form of a question (Chapter 2). The inquiry then proceeds with the formulation of one or several hypotheses as possible solutions to the problem (Chapter 3). In the next phase of the scientific investigation the hypothesis (or the hypotheses if there are more than one) is tested to determine whether it is probably true or false. This amounts to conducting a study in which certain empirical results that relate to the hypothesis are obtained. The results of the study are then summarized in the form of an *evidence report*. For our present purposes we shall simply consider that an evidence report is a summary statement of the results of an investigation, i.e., it is a sentence which concisely states what was found in the inquiry. Once the evidence report has been formed it is related to the hypothesis. By comparing the hypothesis (the prediction of how the results of the experiment *should* turn out) with the evidence report (the statement of how the results *did* turn out), it is possible to determine whether the hypothesis is probably true or false. We now need to inquire into the various methods in psychology of obtaining data which may be used to arrive at an evidence report. This amounts to an inquiry into the various forms of scientific investigation in psychology.

Methods of Obtaining the Evidence Report

The methods to be discussed have in common the fact that they allow the investigator the opportunity to collect data through which, regardless of the specific method used to obtain them, an evidence report can be formulated.

Nonexperimental Methods

Since this is a book on experimental methodology we should not consider nonexperimental methods to any great extent. But to enhance our understanding of experimental methods let us examine them briefly. The manner of classifying the nonexperimental methods is somewhat arbitrary and varies with the classifying authority. There are two general types which can be contrasted with the experimental method.

The first general type of method to consider is the *clinical method* (sometimes called the "case history method" or the "life history method"). The psychologist uses this method in an attempt to help the individual solve his problems, be they emotional, vocational, or whatever. In the most general form of the clinical method, the psychologist attempts to help by collecting information about the person from every possible source. He is interested in all aspects of a person's life, from birth to the present time. Some of the techniques for collecting this information might be an intensive interview, perusing records, administering psychological tests, questioning other people about the individual, studying written works of the person, or obtaining biographical questionnaires. Then, on the basis of such information the psychologist tries to determine the factors which led to the development of the person's problem. This leads to the formulation of an informal hypothesis as to the cause of the person's problem; and the collection of further data will help him to determine whether the hypothesis is probably true or false. Once the problem, and the factors which led to its development, is laid bare for the person, the psychologist will try to help his subject achieve a better adjustment to the circumstances. It may be noted that the clinical method is generally used in an applied, as against a basic sense, since its usual aim is to solve a practical problem, not to advance science. (See p. 267).

The second general method we might consider is the method of *systematic observation* or the *field study* method. In using this method the investigator goes into the field to collect his data—he takes an event as it occurs naturally and studies it, with no effort to produce or to control the event, as in experimentation. The observation of children at free play would be one example of the use of this method. The purpose there might be to determine what kinds of skills children of a certain age possess. A number of types of play apparatus would be made available for the children, and their behavior would be observed and recorded as they played. Another example of the use of the method of systematic observation might be a study of panic. We do not ordinarily produce panic in groups of people for psychological study. Rather, psychologists usually wait until a panic occurs naturally, and then set out to study it. An example of how social psychologists studied a panic was after the Orson Welles' radio dramatization of H. G. Wells' *War of the Worlds.* Psychologists were interested in why the panic occurred and thus went out and interviewed people who participated in it (Cantril, 1940).

The data obtained by the method of systematic observation can be used for testing hypotheses through the construction of evidence reports, but it should be obvious that rather important limitations exist. Such limitations will be briefly discussed after a consideration of the experimental method.

The Experiment

In the initial stages of a science, or some aspect of that science, the nonexperimental methods are used. In some sciences, sociology, for example, there is little hope that anything but nonexperimental methods can ever be used. This is primarily because sociology is largely concerned with the effects of the prevailing culture and social institutions on behavior, and it would be difficult to manipulate these two factors as independent variables in an experiment. In those fields that are ultimately susceptible to experimentation, however, a change in methodology eventually occurs as knowledge accumulates. Hull (1943, p. 1) points out that as scientific investigations become more and more searching, the "spontaneous" happenings in nature are not adequate to permit the necessary observations. This leads to the setting up of special conditions to bring about the desired events under circumstances favorable for scientific observations—and experiments thus originate. Thus, the experimenter takes an active part in producing the event; some advantages of this approach are well expressed by Woodworth and Schlossburg (1955, p. 2):

1. The experimenter can make the event occur when he wishes. So he can be fully *prepared* for accurate observation.
2. He can *repeat* his observation under the same conditions for verification; and he can describe his conditions and enable other experimenters to duplicate them and make an independent check of his results.
3. He can *vary the conditions* systematically and note the variation in results . . .

Since psychology is concerned with the behavior of organisms, in using nonexperimental methods the psychologist must wait until the behavior in which he is interested occurs naturally. The researcher does not have control over the variable that he wishes to study; he can only observe it in its natural state. The one characteristic that all the nonexperimental methods have in common is that the variables that are being evaluated are not purposefully manipulated by the researcher.[1]

[1] To emphasize this definition of an experiment, suppose that a researcher is interested in the way that learning speed changes with age. He might have a group of 20-year-old people and a group of 60-year-old people. Both groups would learn the same task. While at first glance this might appear to be an "experiment," we would not so classify it because the researcher did not *purposely manipulate* his "independent variable" (age of the subjects). Rather, he *selected* his subjects to fit certain age requirements. It is apparent that "age of subjects" is not a variable over

It follows that when a theory is tested through the use of the experimental method, the conclusion is more highly regarded than if it is tested by a nonexperimental method. Put another way, the evidence report which is obtained through experimentation is more reliable than that obtained through the use of a nonexperimental method. This is so, largely because the interpretation of the results is clearer, in an experiment. Ambiguous interpretation of results can occur in nonexperimental methods primarily because of a lack of control over extraneous variables. It is difficult, or frequently impossible, in a systematic observation study to be sure that the findings, with respect to the dependent variable, are due to the independent variable, for they may result from some uncontrolled extraneous variable which happened to be present in the study. In nonexperimental methods it is also usually more difficult to define the variables studied than where they are actually produced, as in an experiment.

All of this does not mean, however, that the experimental method is a perfect method for answering questions. Certainly it can lead to errors and in the hands of poor experimenters the errors are sometimes great. Relatively speaking, however, the experimental method is preferred *where it can be appropriately used.* But if it cannot be used, then we must do the next best thing, and use the method of systematic observation. Thus when it is not possible to produce the events that we wish to study, as in the panic example, we must rely on nonexperimental methods. But we must not forget that when events are *selected* for study, rather than being *produced* and *controlled,* caution must be used in accepting the result.

One frequent criticism of the experimental method is that when an event is brought into the laboratory for study (as it usually is), the nature of the event is thereby changed. For one thing, the event does not naturally occur in isolation, as it is made to occur (relatively so) in the laboratory, for in natural life there are always many other variables which influence it. Criticism of experiments on such grounds is unjustifiable, since we really want to know what the event is like when it *is* uninfluenced by other events. It is then possible to transfer the event back to its natural situation at which time we know more about how it is produced. The fact that an event may appear to be different in the natural situation, as compared to the laboratory, simply means that it is being influenced by other variables, which *in their turn* need to be brought into the laboratory for investigation. Once all the relevant variables which exist under natural conditions have been studied in isolation in the laboratory, and it has been determined in what way they all in-

---

which a researcher has control—he cannot say to one subject: "You will be 20 years old," and to another subject: "You will be 60 years old." Hence, in this investigation the method of systematic observation was used.

fluence the dependent variable and each other, then a thorough understanding of the natural event will have been accomplished. Without such an analysis of events in the laboratory, it is likely that we would never be able adequately to understand them.

Now, while we have made some strong statements in favor of the laboratory analysis of events, we must recognize the possibility that they actually may be changed or even "destroyed." This occurs when the experimenter has simply not been successful in transferring the event into the laboratory. He has produced a truly different event than that which he wished to study. For instance, it may be that if a subject knows that he is in an experiment, he may actually act differently than he would in "real life." The best that can be said for this situation is that usually events cannot be studied adequately unless they *are* brought into this laboratory and at least in the laboratory suitable controls can be introduced so that even if the event is distorted by observation, at least the event is studied with this effect held constant.

Skinner (1953, p. 435) makes a similar point when he says, in essence, that certain characteristics of behavior are too complex to understand through casual observation of everyday life—to find the relevant variables that determine a certain kind of behavior would require considerable time, and even then it may not be feasible to obtain through such a common sense approach. But with adequate recording devices in the laboratory, and under controlled conditions, we may determine the variables that are responsible for an event. These findings can be utilized to good advantage in further study of the complex world at large. To illustrate his point, Skinner suggests that casual observation has led us to conclude that a particularly undesirable type of behavior may be eliminated by punishing a person. Punishment gives quick results; a particular type of behavior seems to disappear immediately if its perpetrator is punished. However, actual laboratory studies have indicated that punishment is rather ineffective in eliminating a response. While it may effect a temporary suppression, it does not seem to permanently eliminate it, even when it is applied over a long period of time. A more effective technique for eliminating a response is the process of extinction, but the discovery of this finding from the observation of everyday life required laboratory investigation.

Types of Experiments

In your reading you may run across a number of terms which refer to different types of experiments; to prevent possible confusion some of them will be discussed here. First, note well that the *general experimental method* remains the same, regardless of the type of problem to which it is applied, so that, strictly speaking, we are not talking about different types of experiments, but different types of problems and purposes for which they are used. To clarify this matter, contrast *exploratory* with *confirmatory* experiments. The type the experimenter uses depends on the state of the knowledge relevant to the problem with which he is dealing. If there is little knowledge about a given problem, the experimenter performs an exploratory experiment. Lacking much knowledge about the problem he is usually not in a position to formulate a possible solution—generally, he cannot postulate an explicit hypothesis which might guide him to predict such and such a happening. He is simply curious, collects some data, but doesn't really have any basis on which to guess how the experiment will turn out. Although he does not have an explicit hypothesis, he evidently has an "informal" hypothesis, at least to the extent that he has decided to investigate the effect of one specific variable on another, rather than any variable to a host of other variables. But his hypothesis is not sufficiently advanced to say what kind of effect the one variable will have on the other, or even to say that there *will* be an effect. It can be seen that the exploratory experiment is performed in the earlier stages of the investigation of a problem area. As he gathers data relevant to the problem, the experimenter becomes increasingly capable of formulating hypotheses of a more clear-cut nature—he is able to predict, on the basis of a hypothesis, that such and such an event should occur. At this stage of knowledge development he performs the confirmatory experiment, i.e., he starts with an explicit hypothesis that he wishes to test. On the basis of that hypothesis he is able to predict an outcome of his experiment; he sets up the experiment to determine whether the outcome is, indeed, that predicted by his hypothesis. Put another way, in the exploratory experiment the scientist is interested primarily in finding new independent variables that affect a given dependent variable, while in the confirmatory experiment he is interested in confirming that a given variable is influential. In the confirmatory experiment he may also want to determine the extent and precise way in which one variable influences the other, or more generally, to determine the functional (quantitative) relationship between the two variables. Underwood (1949, pp. 11-14) has used two descriptive terms to refer to the problems which the two types of experiments are used to solve. The exploratory experiment refers to his "I-wonder-what-would-happen-if-I-did-this" type of problem, while the confirmatory experiment is analogous to his "I'll-bet-this-would-happen-if-I-did-this" type of problem.

But regardless of the state of knowledge in a given problem area, the immediate purpose of an experiment is to arrive at an evidence report.

If the experiment is exploratory, the evidence report can be used as the basis for formulating a specific, precise hypothesis. In a confirmatory experiment the evidence report is used to determine whether the hypothesis is probably true or false. In the latter case, if the hypothesis is not in accord with the evidence report, it might be modified to better fit the data and then tested in a new experiment. If the hypothesis is supported by the evidence report, then its probability of being true is increased. In addition to providing such a general understanding, the distinction between these two types of experiments has direct implications for the types of experimental designs that are employed, one type of design being more efficient for the exploratory experiment, while another is more efficient for the confirmatory experiment.

Sometimes you may run across the term "crucial experiment" (or *experimentum crucius*). This term is used to describe an experiment which purports to test one or several "counter-hypotheses" simultaneously. For instance, it may be possible to design the experiment in such a manner that if the results come out one way, one hypothesis can be said to be confirmed and a second hypothesis disconfirmed, while if the results point another way, the first hypothesis is said to be disconfirmed and the second hypothesis is confirmed. Ideally, a crucial experiment is one whose results support one theory, and disconfirm all possible alternative theories. However, we can seldom if ever be sure that we can state at any given time all of the possible alternative hypotheses. Accordingly, perhaps we can never have a crucial experiment, but only approximations of one.

The term *pilot study* or *pilot experiment* has nothing to do with the behavior of aircraft operators, as one student thought, but refers to a preliminary experiment, one conducted prior to the final experiment. It is used, usually with only a small number of subjects, to suggest what specific values should be assigned to the variables being studied, to try out certain procedures to see how well they work, and more generally to find out what mistakes might be made in conducting the actual experiment so that the experimenter can be ready for them. It is a "dress rehearsal" of the final experiment.

Planning an Experiment

Given a problem to be solved, and a hypothesis, which is a tentative solution to that problem, we must design an experiment which will determine whether that hypothesis is probably true or false. In designing an experiment the researcher uses his ingenuity to obtain data that are relevant to the hypothesis. This involves such problems of experimental technique as: What apparatus will best allow manipulation and observation of the phenomenon of interest? What extraneous variables may contaminate the primary phenomenon of interest and are therefore in need of control? Which events should be observed and which should be ignored? How can the results of the experiment best be observed and recorded? By considering these and similar problems an attempt is made to rule out the possibility of collecting irrelevant evidence. For instance, if the antecedent conditions of the hypothesis are not satisfied, the evidence report will be irrelevant to the hypothesis, and further progress in the inquiry is prohibited. Put another way, the hypothesis says that *if* such and such is the case (the antecedent conditions of the hypothesis), *then* such and such should happen (the consequent conditions of the hypothesis). The hypothesis amounts to a contract that the experimenter has signed—he has agreed to make sure that the antecedent conditions are fulfilled. If he fails to fulfill his agreement, then whatever results he collects will have nothing to do with his hypothesis—they will be irrelevant and thus cannot be used to test the truth of the hypothesis. This points up the importance of adequately planning the experiment. For if the experiment is improperly designed, then either no inferences can be made from it, or it may only be possible to make inferences to answer questions that the experimenter has not asked. That this is not an idle warning is indicated by the frequency with which the right answer is given to the wrong question, particularly by neophyte experimenters. And if the only result of an experiment is that the experimenter learns that he should not make these same errors in the future, this is very expensive education, indeed.

It is a good idea for the experimenter to draft a rather thorough plan of the experiment before he conducts it, although this is not always necessary for more sophisticated researchers. Once the complete plan of the experiment is set down on paper it is desirable to obtain as much criticism of it as possible. The experimenter often overlooks many important points, or looks at them with a wrong, preconceived set, and the critical review of others may bring potential errors to the surface. No scientist is beyond criticism, and it is far better for him to accept criticism before an experiment is conducted, than to make errors that might invalidate the experiment. We shall now suggest a series of steps which the experimenter can follow in the planning of an experiment.

Outline for an Experimental Plan

1. *Label the Experiment.* The title should be clearly specified, as well as the time and location of the experiment. As time passes and the experimenter accumulates a number of experiments, he can always refer to this

information without much chance of confusing one experiment with another.

2. *Survey of the Literature.* All of the previous work that is relevant to the experiment should be studied. This is a particularly important phase in the experimental plan for a number of reasons. First, it helps in the formulation of the problem. The experimenter's vague notion of what problem he wants to investigate is frequently made more concrete by consulting other studies. Or, the experimenter thus may be led to modify his original problem in such a way that the experiment becomes more valuable. Another reason for this survey of pertinent knowledge is that it tells the experimenter whether or not the experiment even needs to be conducted. If essentially the same experiment has previously been conducted by somebody else, there is certainly no point in repeating the operation, unless it is specifically designed to confirm previous findings. Other studies in the same area also provide numerous suggestions about extraneous variables that need to be controlled and hints on how to control them.

The importance of the literature survey cannot be overemphasized. The experimenters who tend to slight it usually pay a penalty in the form of errors in the design or some other complication. The knowledge in psychology is growing all the time, making it more difficult for one person to comprehend the findings in any given problem area. Therefore, this step requires particularly close attention. Also, since reference to relevant studies should be made in the write up of the experiment anyhow, this might just as well be done before the experiment is conducted thus combining two steps in one.

We are very fortunate in psychology to have the *Psychological Abstracts* which makes any such survey relatively easy.[2] Every student of psychology should attempt to develop a facility in using the *Abstracts*.

3. *Statement of the Problem.* The experiment is being conducted because there is a lack of knowledge about something. The statement of the problem expresses this lack of knowledge. While the problem can be developed in some detail, through a series of logical steps, the actual statement of the experimental question should be concise—it should be stated succinctly and unambiguously in a single sentence, preferably as a question. The statement of the problem as a question implies that it can be answered unambiguously in either a positive or negative manner. If the question cannot be so answered, *in general* we can say that the experiment should not be conducted. But every worthwhile experiment in-

---

[2] The Psychological Abstracts is a professional journal published bimonthly by the American Psychological Association. It summarizes the large majority of psychological research and classifies it according to topics (and authors) so that it is fairly easy to determine what has previously been done on any given problem.

volves a gamble. If the problem cannot be definitely answered either positively or negatively, the experimenter has not risked anything and therefore cannot hope to gain new knowledge.

4. *Statement of the Hypothesis.* The variables specified in the statement of the problem are explicitly stated in the hypothesis as a sentence. Natural languages (e.g., English) are usually employed for this purpose, but other languages (e.g., mathematical or logical ones) can also be used, and, in fact, are preferable. The "if . . . then . . ." relationship was suggested as the basic form for stating hypotheses.

5. *Definition of Variables.* The independent and dependent variables have been specified in the statement of the problem and of the hypothesis. They must now be defined in such a manner that they are clear and unambiguous. This is particularly important in scientific research and for this reason it is desirable to consider it in some detail.

There are many terms that are used both in science and in everyday life whose definitions are ambiguous. Also, many terms are defined in several different ways. Everyone can recall, no doubt, at least several lengthy and perhaps heated arguments which, on more sober reflection, were found to have resulted from a lack of agreement on the definition of certain terms that were basic to the discussion. To illustrate, suppose a group of people carried on a discussion about happiness. The discussion would no doubt take many turns, produce many disagreements, and probably result in considerable unhappiness on the part of the disputants. It would probably accomplish little, unless at some early stage in the discussion the people involved were able to agree on an unambiguous definition of "happiness." While it is impossible to guarantee the success of a discussion in which the terms are adequately defined, *without* such an agreement there would be no chance of success whatsoever. The implications of this requirement for science should be clear—the terms used must be adequately defined.

The main functions of adequate definitions in science are (1) to clarify the phenomenon being considered and (2) to allow us to communicate with each other in an unambiguous manner. And the type of definition with which we are mainly concerned in experimentation is the *operational definition.* Essentially, an operational definition is one that indicates that a certain phenomenon exists and does so by specifying precisely how (and preferably in what units) the phenomenon is measured. That is, an operational definition of a certain concept consists of a statement of the operations necessary to produce the phenomenon. By specifying the operations involved in producing the phenomenon, that phenomenon is said to be operationally defined. To specify precisely the defining operations obviously accomplishes the main intent of the scientist—it produces a phenomenon that a number of observers can agree on, one that can be observed directly. Furthermore, a phenomenon that is operationally defined is reproducible by other people. In short when we operationally define a concept we do so in terms of objective operations; and the definition of the concept consists of the operations performed in arriving at it. For example, when we define air temperature, we mean that the column of mercury in a thermometer rests at a certain point on the scale of degrees. Consider the psychological concept of hunger drive. One way of operationally defining this concept is in terms of the amount of time that an organism is deprived of food. Thus, an operational definition of *hunger drive* would be a statement about the number of hours of food deprivation. Accordingly, we might say that an organism that has not eaten for twelve hours is more hungry than one which has not eaten for two hours.

A considerable amount of work has been done in psychology on steadiness. There are a number of different ways of measuring steadiness, and accordingly there would be a number of different operational definitions of the concept. For example, let us consider the Whipple Steadiness Test (see Figure 4.1). The apparatus in this test consists of a series of holes,

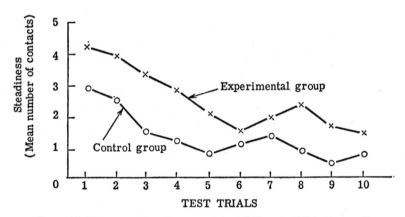

Figure 4.1. Mean number of contacts per trial on the Whipple steadiness test. The smaller the number of contacts, the greater the steadiness.

varying from large to small in size, and a stylus. The subject attempts to hold the stylus as steady as he can in each hole, one at a time, without touching the sides. The number of contacts which he makes is automatically recorded. Presumably, the steadier subject will make fewer contacts than the unsteady subject. Accordingly, we would operationally define steadiness as the number of contacts made by a subject when taking the Whipple Steadiness Test. Let us add that if we measured steadiness by using other types of apparatus, then we would have additional operational definitions of steadiness.

The value of adequate operational definitions cannot be overemphasized, for every term in a hypothesis must be capable of operational definition. This amounts to saying that for every term in a hypothesis we must be able to refer to (to "point" to) some corresponding event in the environment. If no such operation is possible for all the terms in the hypothesis, we must conclude that the hypothesis is meaningless.[3]

6. *Apparatus.* Every experiment involves two things: (1) an independent variable must be manipulated; and (2) the resulting value of the dependent variable must be recorded. Perhaps the most frequently occurring type of independent variable in psychology is the presentation of certain values of a stimulus; and in every experiment a response is recorded. Both of these functions may be performed manually by the experimenter. However, it is frequently desirable, and in fact sometimes necessary to resort to mechanical or electrical assistance. We may even make the bold statement that the more an experimenter can rely on apparatus in his experiment, the better off he will be. There are naturally some exceptions to this general statement, but if the apparatus is adequate for the job, those exceptions will be few. Thus, there are two general functions of apparatus in psychological experimentation: (1) to facilitate the administration of the experimental treatment; and (2) to aid in recording the resulting behavior. Let us consider how these two functions might be advantageously accomplished in an experiment.

Some interest has been shown in the use of subliminal stimulation in advertising. Let us assume that the Chestostrike Cigarette Company becomes interested in this technique, but they want experimental proof that it will work before they begin using it on a large scale. An experimental psychologist conducts the following experiment for them. Two groups of subjects watch a television program. While the experimental group is watching, however, the words "Smoke Chestostrike Cigarettes" are flashed on the screen every two minutes. But, since we are concerned with subliminal stimulation, the stimulation provided by the words is below the subjects' threshold—the words are flashed, but for such a short time that

---

[3] In 1929 P. W. Bridgeman's book (1927) was published, initiating the movement known as operationism. The prime assumption of operationism is that the adequate definition of the variables with which a science deals is a prerequisite to advancement. In the years that have followed much has been written concerning operationism; these writings have led to many polemics. We have chosen to avoid becoming involved in such problems, but it is worthwhile to point out that they exist. The student should thus realize that a discussion of operational definitions can lead into matters far beyond what we have taken up here. For some of the more philosophical discussions of this topic consult Frank (1956), while for a more extended consideration of how operational definitions are used in psychology read the excellent discussion in Underwood (1957).

the subjects are not aware that they are reading anything (they cannot report that any words are being flashed on the screen). The control subjects, on the other hand, simply watch the program without the subliminal stimulation. When the program is over the experimenter may ask the subjects to fill out a brief form indicating how well they liked the program (a ruse to distract their attention from the main purpose of the experiment). The experimenter then tells the subjects that he appreciates their cooperation and says that they are to be compensated for their time with a package of cigarettes. The subjects then select whatever brand they want out of several, including Chestostrikes, displayed in a cigarette machine. The results show that more Chestostrikes are selected by the experimental group than by the control group. It is thus concluded that this type of subliminal stimulation is effective.

Let us now return to the main point we wish to illustrate through this fictitious experiment. We may note that apparatus was used to present the experimental treatment—in fact, in this experiment it was essential. The particular type of apparatus used was a tachistoscope—a device for presenting a stimulus for whatever length of exposure an experimenter may desire. In this case the tachistoscope was used to present the image "Smoke Chestostrike Cigarettes" on the television screen. In our situation the image had to be presented for an extremely short period of time, perhaps 1/200 of a second. Furthermore, there was an automatic timing device attached to the tachistoscope with a driving motor so that the words were exposed at precisely two-minute intervals. The second function of apparatus in experimentation was fulfilled by the cigarette machine. By allowing the subject to push one of several buttons, the particular brand of cigarettes that he desired was delivered. And we may assume that an automatic counter was attached to the delivery mechanism in order to allow the experimenter to ascertain the number of packages of each brand selected.

The types of apparatus used in experimentation are so numerous that we cannot attempt a systematic coverage of them here.[4] We shall, however, briefly try to illustrate further the value of apparatus, as well as to refer to certain cautions that should be observed.

Frequently a certain stimulus, such as a light, must be presented to a subject at very short intervals. It would be difficult for an experimenter to time the intervals manually and thus make the light come on at precisely the desired moments. In addition to the considerable error that would be involved, the work required of the experimenter would be highly undesirable, and in fact might distract him from other important aspects of the experiment. This type of stimulus presentation could be handled

---

[4] An excellent general source is the book by Grings (1954); for a more detailed presentation of electronic instrumentation see Brown & Saucer (1958).

very easily by placing a metronome in the circuit to break and complete
the circuit at the proper times. As you can see, one of the main advantages
of apparatus is that it reduces the "personal equation." Suppose that a
reaction-time experiment is being conducted. For example, one might
wish to present a series of words, one at a time, to a subject and measure
the time it takes him to respond with the first word that comes to mind.
If the experimenter were forced to measure the subject's reaction time by
starting a stop watch when the word is read and stopping it when the
subject responds, considerable error would result. The reaction time of
the experimenter in starting and stopping the stop watch enters into the
"subject's reaction time"; this is especially bad because the experimenter's
reaction time would be bound to show considerable variation. A better
approach would be to use a voice key that is connected in circuit with a
timing device. In this apparatus, when the word is read to the subject,
the timer automatically starts, and when the subject responds, it automatically stops. The experimenter may then record the reaction time and
say the next word. While the "reaction time" of the apparatus is still
involved, it is at least constant and of smaller magnitude, as compared to
the experimenter's reaction time. A similar example of the valuable use of
apparatus would be timing a rat as he runs a maze. Several types of
apparatus that automatically record the rat's latency and running time
have been developed for this problem.

While we have been emphasizing the *desirability* of apparatus in experimentation, there are a large number of problems that simply cannot
be studied without the use of apparatus. In some experiments apparatus
is not only valuable, it is downright essential. Probably the clearest
illustration would be the recording of response measures of a physiological nature. If one wishes to study the effect of some independent
variable on electroencephalograms ("brainwaves"), an electroencephalo-
graph is required. Similarly, for measuring pulse rate a sphygmograph is
necessary, for recording galvanic skin responses a psychogalvanometer
needs to be used, and so forth.

But in spite of all the advantages of apparatus, there are a number of
possible disadvantages. Some of these disadvantages can better be discussed in the chapter on control (Chapter 6). Suffice it to say here that
we have assumed above that, to be of value, the apparatus is suitable for
the job required of it. This is not always the case, for sometimes apparatus
is not accurate, not adequately calibrated, etc. Furthermore, sometimes
apparatus may interfere with the event being studied.

7. *Control of Variables.* In this phase of planning the experiment the
scientist must consider all of the variables that might contaminate the
experiment—he should attempt to evaluate any and all variables which
might affect his dependent variable. He may decide that some of these

extraneous variables might act in such a way that they will invalidate the experiment, or at least leave the conclusion of the experiment open to question. Such variables need to be controlled. To elaborate on what we mean by "control" and the techniques for achieving it will demand much attention; since it is of extreme importance in experimentation, it will be the subject of a later discussion (Chapter 6). For now, let us simply note a point which will be expanded later—that we must make sure that no extraneous variable will differentially affect our groups, i.e., that no such variable will affect one group differently than it will affect another group.

In the course of examining variables that may influence our dependent variable we will have to make certain decisions. We will decide, preferably on the basis of information afforded by previous research, that a certain variable is not likely to be influential, in which case its effects can be fairly safely ignored. We need not control such variables.

Other variables, however, might be considered to be relevant, but we do not know how to control them. Perhaps it is reasonable to assume that such variables will exert equal effects, at random, on all conditions. If this assumption seems tenable, the experimenter might choose to proceed. But if such an assumption is rather tenuous, then the problem may well be of such serious proportions that it would be wiser to abandon the experiment.

8. *Selection of a Design.* So far we have discussed only the two-groups design, i.e., where the results for an experimental group are compared to those for a control group. We shall consider a number of other designs later, from among which the experimenter may choose the one most appropriate to his problem. For example, it may be more advantageous to use several groups, instead of just two, in which case a *multi-groups* design would be adopted. Another type of design, which in many cases is the most efficient and which is being used more and more in psychology, is the *factorial* design. For now, however, the reader should simply consider the two-groups design, at the same time realizing that a number of alternatives are possible.

9. *Selection and Assignment of Subjects to Groups.* The experimenter usually conducts an experiment because he wants to conclude something about the behavior of a group of subjects. To do this, of course, he must select certain subjects to study. But from what larger group should he select his subjects? This is an important question because he will want to generalize his findings from the subjects that he is studying to the larger group of subjects from which they were chosen (see step 14, page 63). The larger group of subjects is the *population* (or "universe") under study; those that participate in his experiment constitute his *sample*. More generally, by "population" we mean the total number of possible items of a class that might be studied—it is the entire group of

items from which a sample has been taken. Thus, we may note that a population need not refer only to people, but to any type of organism: amoebae, rats, jellyfish, etc. Furthermore, one may note that the definition is worded in such a manner that it can and does refer to inanimate objects. For example, we may have a "population" of types of therapy (directive, nondirective, etc.), of learning tasks (hitting a baseball, learning a maze, etc.), and so forth. Or an experimenter may be interested in sampling a population of stimulus conditions (high, medium, low intensity of a light), or a population of experiments (three separate experiments that test the same hypotheses, etc.). In engineering, a person may be interested in studying a population of the bridges of the world, and an industrial psychologist may be concerned with a population of whiskey products, and so on.

In designing an experiment one should specify with great precision the population (or populations) he is studying. For the moment let us merely concern ourselves with subject populations, and leave other populations until later (see Chapter 13). In specifying a population we must note those of its characteristics that are particularly relevant to its definition, e.g., if we are concerned with a population of people, we might wish to specify the age, sex, education, socio-economic status, race, etc. If we are working with animals, we might wish to specify the species, sex, age, strain, experience, habitation procedures, feeding schedules, etc. Unfortunately, one can observe by reading articles in professional journals that experimenters rarely define the populations that they are studying with any great precision.

Given a well defined population which we wish to study, then, we are faced with the problem of how to accomplish this study. If it is a small population, it may be possible to study each individual. And adequately studying an entire population is far preferable to studying a sample of it. The author once conducted some consumer research studies in which he was supposed to obtain a sample of eighteen people in a small town in the High Sierras. After considerable difficulty he was able to locate the "town" and after further difficulty he was eventually able to find an eighteenth person. In this study, the entire population of the town was exhausted, and more reliable results were obtained than if a smaller sample of the town was selected. As it turned out, however, not a single person planned on purchasing TV sets, dishwashers, or similar electrical appliances during the next year, largely because the town did not have electricity.

But in any event, the population to be studied is seldom so small that it can be exhausted by the researcher. More likely, the population is so large, sometimes infinitely large, that it cannot be studied in its entirety, and the researcher must resort to studying a sample. One of the reasons that a population may be infinitely large (or perhaps finite, but indefinitely large), is that the experimenter may wish to generalize not only to people now living, but to people who are as yet unborn.

Where the population is too large to be studied in its entirety, the experimenter must select a number of subjects and study them. One technique that an experimenter may use in selecting a sample is that of *randomization*. In random selection of a sample of subjects from a population each member of the population has an equal chance of being chosen. For instance, if we wish to draw a random sample from a college of six hundred students, we might write the names of all the students on separate pieces of paper. We would then place the six hundred slips of paper in a hat, mix thoroughly, and, without looking, draw our sample. If our sample is to consist of sixty students, we would select sixty pieces of paper. Of course there are simpler techniques to achieve a random sample. For instance, we might take a list of all six hundred students and select every tenth one to form our sample.

Once the experimenter randomly selects a sample, he then assumes that his sample is typical of the entire population—that he has drawn a *representative* sample. Drawing samples at random is usually sufficient to assure that his sample is representative, but the researcher may check on this if he wishes.[5] For one, if he has values available for the population, then he can compare his sample with the population values. If he is studying the population of people in the United States, he has readily available a large amount of census information about education levels, age, sex, etc.[6] He can compute certain of these statistics for his sample, and compare these figures with those for general population. If they jibe, he can assume that his sample is representative. The assumption that he is making, in this case, is that if his sample is similar to the population in a number of known characteristics, it is also similar with respect to characteristics for which no data are as yet available. This could be a dangerous assumption, but it is certainly better than if there was no check on representativeness.

---

[5] A somewhat infrequent definition of a representative sample is one that has been randomly drawn from a population. Following this definition, of course, it would be foolish to check the representativeness of the sample. We are, however, following the more common procedure of assuming that a sample that is not representative (is very atypical) of a population can be randomly selected. For example let us define a population of people that consists of 90 blonds and 10 redheads. Now if we draw a random sample of 10, and find that they are all redheads, we would say that the sample is not representative of the population. See Lindquist (1953) for an excellent discussion of this topic.

[6] Of course, some census statistics are also obtained through sampling techniques and such statistics do not guarantee that the population has those precise values. We are simply assuming here that such statistics are *probably* true—a rather safe assumption.

Once the population has been specified, a sample drawn from it, and the type of design determined, it is necessary to divide the sample into the number of groups to be used. The subjects must be assigned to groups by some random procedure (assuming that what we shall later call a matched groups design is not used). By using randomization we would assure ourselves that each subject has an equal opportunity to be assigned to each group. Some measure such as coin flipping can be used for this purpose. For example, suppose that we have a sample of twenty subjects, and that we have two groups. We might take the first subject and flip a coin; if it is "heads," the subject will be placed in group 1, and if it is "tails," he will be placed in group 2. We would then do likewise for the second subject, and so on until we have ten subjects in one group. The remaining subjects would then be assigned to the other group.

We now have two groups of subjects who have been assigned at random. The next thing to do is to determine which group is to be the experimental group and which is to be the control group. This decision should also be determined in a random manner, such as by flipping a coin. We might make a rule that if a "head" comes up, group 1 is the experimental group and group 2 the control group; but if our coin flipping yields a "tail," group 1 would be the control group, group 2 the experimental group.

By now you should have acquired a feel for the importance of randomization in experimental research—the random selection of a sample of subjects from a population, the random assignment of subjects to groups, and the random determination of which of the two groups will be the experimental group and which will be the control group. It is by the process of randomization that we attempt to eliminate biases (errors) in our experiment. When we want to make statements about our population of subjects, we generally study a sample that is representative of that population. If our sample is not representative, then what is true of our sample may not be true of our population, and we might make an error in generalizing the results obtained from our sample to the population. Random assignment of our sample to two groups is important because we want to start our experiment with groups that are essentially equal. If we do not randomly assign subjects to two groups, we may well end with two groups that are unequal in some important respect. If we assign subjects to groups in a nonrandom manner perhaps just looking at each subject and saying "I'll put you in the control group," we may have one group being more intelligent than the other—consciously or unconsciously, we may have selected the more intelligent subjects for the experimental group.

Random assignment of subjects to groups helps prevent biases from entering into the experiment. In short, randomization is used largely to prevent any biases in favor of one group over the other. We must hasten to add that the use of randomization does not guarantee that our sample is representative of the population from whence it came, or that the groups in an experiment are equal before the administration of the experimental treatment. For, by chance, randomization may produce two unequal groups—one group may in fact turn out to be significantly more intelligent than the other. However, randomization is frequently the best procedure that we can use, and we can be sure that, at least in the long run, its use is justified. For any given sample, or in any given experiment, randomization may well result in "errors," but here, as everywhere else in life, we must play the probabilities. If a very unlikely event occurs (e.g., if the procedure of randomization leads to two unequal groups) we will end up with an erroneous conclusion. Eventually, however, due to the self-checking nature of science, the error will be discovered.

We may conclude discussion of this step of the experimental plan by noting that the number of groups to be used in an experiment is determined by the number of independent variables and the number of values of them that we have selected for study, as well as by the nature of the variables to be controlled. For instance, if we have a single independent variable that we decide to vary in two ways, we would have two groups. More than likely these two groups would be called experimental and control groups. If we select three values of the independent variable for study, then obviously we would need to assign our sample of subjects to three groups. It might be added that usually an equal number of subjects is assigned to each group. Thus if we have eighty subjects in our sample, and if we vary the independent variable in four ways, we would have four groups in the experiment, probably twenty subjects to each group. It is not necessary, however, to have the same number of subjects in each group, and the experimenter may determine the size of each group as he sees fit.

10. *Experimental Procedure.* The procedure for conducting the data collection phase of the experiment should be set down in great detail. The experimenter should carefully plan how the subjects will be treated, how the stimuli will be administered, and how the response will be observed, and recorded, etc. He should specify the instructions to the subjects (if humans are used) and formulate a statement concerning the administration of the independent variable and the recording of the dependent variable. It might be useful if the experimenter makes an outline of each point to be covered in the actual data collection phase. He might start his outline right from his greeting to the subject and carry through step by step to when he says goodbye. It is also advisable to try out a few subjects to see how the procedure works. More often than not such "dress rehearsals" will suggest new points to be covered and modifications

of procedures already set down. And if more elaborate checking of the procedure is desired a pilot study might be conducted (see p. 49).

11. *Statistical Treatment of the Data.* The data of the experiment are usually subjected to statistical analysis. As psychology has progressed this phase of experimentation has become increasingly important. We have witnessed the development of some very powerful statistical techniques. In some manner the results of the experiment should be evaluated to determine whether they might have occurred by chance. Suppose, for instance, a two-groups design is used and the mean for the experimental group is found to be 14.0, the mean for the control group 12.1. In this case one might conclude that the experimental group is superior to the control group (if a high score is "good"). On the basis of this limited amount of information, however, such a conclusion is not justified, since it has not been determined whether the difference is a reliable (significant) difference. It may not be a "real" difference, but merely one which has occurred by chance. If the difference is not significant, the next time the experiment is conducted the outcome may be reversed. Thus, as we noticed in Chapter 1, we need to use a statistical technique to determine whether the difference between the mean scores of the two groups is significant. The statistical analysis will tell you the odds that the difference between the groups might have occurred by chance. If the probability that this difference could have occurred by chance is small, then we may conclude that the difference is significant, that the experimental group is reliably superior to the control group.

The main point here is that the data must be evaluated by a statistical test, a number of which are available. Some tests are appropriate to one kind of data or experimental design and some are not. The use of such statistical tests frequently requires that certain assumptions about the experimental design and the kind of data collected must be met. In taking such matters into consideration it is advisable to plan the complete procedure for statistical analysis prior to conducting the experiment. Sometimes experimenters do not do this, and find that there are serious problems in the statistical analysis which could have been prevented by a little more insight. Lack of rigor in the use of statistics can invalidate the experiment. And the statistics used must be appropriate to the design selected.

12. *Forming the Evidence Report.* The evidence report is a summary statement of the findings of the experiment, and will be discussed in detail later (see Chapter 11). These paragraphs will serve merely as a brief survey of the subject.

The evidence report should tell us something more than just the results of the experiment. It should also tell us that the antecedent conditions of the hypothesis held (were actually present) in the experiment. More completely, then, the evidence report is a statement that asserts that the antecedent conditions of the hypothesis obtained, and that the consequent conditions specified by the hypothesis were found either to occur or not to occur. If the consequent conditions were found to occur, we may refer to the evidence report as positive, while if they were not, the evidence report is negative. To illustrate, let us return to a hypothesis previously considered (p. 33): "If a teacher praises a student for good reading performance, then the student's reading growth will increase." Now, let us design an experiment to test this hypothesis. The subjects in an experimental group shall be praised each time they exhibit good reading performance. No praise is given to the members of the control group when they show similar performance. Let us assume that the experimental group exhibits a significantly greater increase in reading growth than does the control group. Referring to the hypothesis, we may note that the antecedent condition was satisfied, and that the consequent condition was found to be the case. We may thus formulate our evidence report: "Students were praised by a teacher when they exhibited good reading performance, and they exhibited an increase in reading growth (as compared to the control group)." In short, the evidence report is a sentence that asserts that the antecedent conditions held *and* that the consequent conditions either did or did not hold—it is of the form "*a* and *b*," where *a* stands for the antecedent conditions and *b* for the consequent conditions of the hypothesis.

13. *Making Inferences from the Evidence Report to the Hypothesis.* In this phase the evidence report is related to the hypothesis for the purpose of determining whether the hypothesis is probably true or false. To do this we must make an inference from the evidence report to the hypothesis. Essentially the inference is the following: if the evidence report is positive, the hypothesis is confirmed (the evidence report and the hypothesis coincide—what was predicted to happen by the hypothesis actually happened, as stated by the evidence report). If, however, the evidence report is negative, we may conclude that the hypothesis is disconfirmed. We may add that, while this is essentially the procedure that experimenters follow, there is considerable question about the validity of the inferences involved. This matter will be considered in Chapter 11.

14. *Generalization of the Findings.* The extent to which the results of the experiment can be generalized depend on the extent to which the populations with which the experiment is concerned have been specified, and the extent to which those populations have been represented in the experiment by random sampling. Considering only subject populations again, let us say that the experimenter specified his population as all of (and only) the students at Ivy College. If he has randomly drawn

a sample of subjects from that population, his experimental results may be generalized to that population; he may assert that what was true for this sample is probably true for the whole population. Of course, if the population was not adequately defined, or the sample was not randomly drawn from it, no such generalization can be made, for the results apply only to the sample studied.

We have now covered every major step of experimentation and you should now have a pretty good idea of the individual steps and how they fall into a logical pattern. Our first effort to present the whole picture was in Chapter 1. In Chapters 2 and 3 and in this section we have attempted to enlarge on some of the steps. Thus in Chapter 2 we considered the nature of the problem, and in Chapter 3 we discussed the hypothesis. These two initial phases of planning the experiment were summarized as steps 3 and 4. Next we considered the way in which the variables specified by the hypothesis are (operationally) defined (step 5). The use of apparatus for presenting stimuli and for recording responses was discussed as step 6. The important topic of control was briefly considered (step 7), but will be enlarged on later (Chapter 6). Following this we pointed out that several designs are possible in addition to the two-groups design that we have largely concentrated on (step 8). The ways in which several different experimental designs may be used is the subject of Chapters 5, 8, 9, 10. Next we took up the selection of subjects and assignment of them to groups (step 9). Step 10 consisted of a brief discussion of experimental procedure, while in step 11 we offered a preview of the techniques of statistical analysis that will be more thoroughly covered in connection with the different experimental designs. The formation of the evidence report, and the way in which it is used to test the hypothesis were taken up in steps 12 and 13. A separate chapter will be devoted to the latter topic (Chapter 11). Finally, we briefly considered the problem of generalization (step 14), a topic that will be more elaborately considered in Chapters 12, 13. Of course, as we continue through the book, each of the above points will continue to make their appearance in a variety of places, even though separate chapters will not be devoted to them. As a summary of this section, as well as to facilitate your planning of experiments, we offer the following check list.

1. Label the experiment.
2. Survey the literature.
3. State your problem.
4. State your hypothesis.
5. Define your variables.
6. Specify your apparatus.

7. State the extraneous variables that need to be controlled and the ways in which you will control them.
8. Indicate the manner of selecting your subjects, the way in which they will be assigned to groups, and the number to be in each group.
9. Draw up a plan for your experimental procedure.
10. Specify the type of statistical analysis to be used.
11. State the possible evidence reports. Will the results tell you something about your hypothesis no matter how they come out?
12. Are you clear about the inferences that can be made from the evidence report to the hypothesis?
13. To what extent will you be able to generalize your findings?

Conducting an Experiment—An Example

Consider an illustration of the various points of the experimental plan. A problem with which we might be concerned is the effect of knowledge of results on performance: we wish to know whether informing a person of how well he performs a task will facilitate his learning of that task. The title of this experiment might be "The Effect of Knowledge of Results on Performance." Assume that the literature survey has been made. The problem may then be stated: "What is the effect of knowledge of results on performance?" Our hypothesis may be: "If knowledge of results is furnished to a person, then that person's performance will be facilitated." Note that the statement of the problem and the hypothesis have *implicitly* determined our variables; we now need to make them *explicit*. The task our subjects shall perform is the drawing, while blindfolded, of five-inch lines. The independent variable concerns the amount of knowledge of results furnished the subjects, and we shall vary it from zero to some large amount. A large amount of knowledge of results may be operationally defined as telling the subject whether his line is "too long," "too short," or "right." "Too long," in turn, may be defined as any line $5\frac{1}{4}$ inches or longer, "too short" as any line $4\frac{3}{4}$ inches or shorter, and "right" as any line between $5\frac{1}{4}$ inches and $4\frac{3}{4}$ inches. A zero amount of knowledge of results would be defined as furnishing the subject no information about how long his line is. The dependent variable would be the actual length of the lines that he draws. We might require the subject to draw fifty lines, his proficiency being determined by his total performance (say, the sum total of his deviations from 5-inch lines) on all 50 trials.

The apparatus would consist of a drawing board on which can be affixed ruled paper; a blindfold; and a pencil. The paper needs to be easily movable for each trial, and ruled in such a manner that the experimenter can tell immediately within which of the three intervals (long, short or right) the subject's lines fall.

We have selected two values of the independent variable for study, a positive amount and a zero amount. We thus require two groups, experimental and control. The experimental group will receive knowledge of results, whereas the control group will not. The subject population may be defined as all of the students in the college. From a list of the student body we will randomly select sixty subjects to study.[7] The sixty subjects are then randomly divided into two groups (see p. 67 for the precise manner in which this assignment will be carried out in this illustration). It is then randomly determined that one of the groups is the experimental group, while the other is the control group.

We now need to determine which extraneous variables might influence our dependent variable and therefore need to be controlled. Our general principle concerning control is that both groups should be treated alike in all respects, except as far as the independent variable is concerned (in this case we administer different amounts of knowledge of results). Hence, essentially the same instructions should be read to both groups; a constant "experimental attitude" should be maintained in the presence of both groups—the experimenter should not frown at some of the subjects and be gay or jovial with others. Incidental cues should be eliminated insofar as possible. An experiment might be invalidated if the experimenter has a habit of holding his breath when the subject reaches the five-inch mark. Not only would this be furnishing knowledge of results to an alert control subject, but it would increase the amount of knowledge of results for the experimental subjects.

Is the amount of time between trials an important variable? Previous research would suggest that it is—in general, the longer the time between trials, the better the performance. We therefore will control this variable by holding it constant for all subjects. We shall specify that each subject will wait ten seconds after each response before his hand is returned to the starting point for the next trial. What other extraneous variables might be considered? Perhaps the time of day at which the subjects are run is important—a person might perform better in the morning than in the afternoon or evening. If the experimental group were run in the morning and the control group in the afternoon, then no clear-cut conclusion about

[7] We need to assume that all sixty subjects will cooperate. The fact that this assumption is questionable leads to the widespread practice among experimenters of using sophomore students in introductory psychology classes—such students are quite accessible to psychologists and usually "volunteer" readily. This method of selecting subjects, of course, does not result in a random sample, and thus leads to the question of whether the sample is representative of the population (all the students in the college).

the effectiveness of knowledge of results could be drawn from the data. One control for this time variable might be to run all subjects between 2 P.M. and 4 P.M. But even this might produce differences, since subjects might perform better at 2 P.M. than at 3 P.M. Furthermore, we could probably not run all the subjects within this one hour on the same day, so we might have to conduct the experiment over a period of two weeks. Now, does it make a difference whether subjects are run on the first day or the last day of the two weeks? It may be that examinations are being given concurrent with the first part of the experiment, causing the subjects to be nervous. Then again it may be that people who are tested on Monday perform differently than people tested on Friday.

The problem of how to control the time variable is rather complex. We shall choose the following procedure (see Chapter 6 for an elaboration): let us specify that all subjects will be run between 2 P.M. and 4 P.M. When the first subject reports to the laboratory, we shall flip a coin to determine whether he will be an experimental or a control subject. If it turns out that he is a control subject, the next subject will be assigned to the experimental group. When the third subject reports, we will similarly determine which group he will be assigned to and the fourth subject will be placed in the other group. The rest of the subjects will be similarly assigned to the groups for as many days as the experiment is conducted.

By using this procedure we may rather safely assume that whatever the effects of time differences on the subjects' performance, they will be balanced—they will affect both groups equally. This is so because, in the long run, we can assume that an equal number of subjects from both groups will participate at any given time of the day, and on any particular day of the experiment. The illustrations given should be sufficient to illustrate the control problems involved. Think of some additional variables that should be considered, e.g., do various distracting influences exist such as noise from radiators, people talking, etc.? We can control these to some extent, but not completely. In the case of those that cannot be reasonably controlled, we must make the assumption that they will affect both of our groups equally—they will "randomize out." For instance, there is no reason to think that the various distracting influences will affect one group to a greater extent than the other. After surveying the various extraneous variables, we may conclude that our assumption is justifiable—there are no variables that cannot either be controlled or whose effect, if any, would differentially affect the dependent variable scores of the two groups.

The next step to consider is the experimental procedure. Our plan for this phase might proceed as follows: "After the subject enters the laboratory room and is greeted, he is seated at a table and given the following instructions: 'I want you to draw some straight lines that are five inches long, while you are blindfolded. You are to draw them horizontally like this (experimenter demonstrates by drawing a horizontal line in the air). When you have completed your line, leave your pencil at the point where you stopped. I shall return your hand to the starting point. Also keep your arm and hand off the table while drawing your line. You are to have only the point of the pencil touching the paper. Are there any questions?' The experimenter will answer any questions by repeating pertinent parts of the instructions. When the subject is ready, the experimenter blindfolds him ("now I am going to blindfold you"), uncovers the apparatus, and places the pencil in the subject's hand. The subject's hand is guided to the starting point and he is instructed: 'Ready? Go.' The experimental subjects are given the appropriate knowledge of results (as previously specified) immediately after their pencils stop. No information is given to the control subjects. After the subject completes a trial, he waits ten seconds, after which his hand is returned to the starting point. He is told: 'Now draw another line five inches long. Ready? Go.' This same procedure is followed until the subject has drawn fifty lines. The experimenter must move the paper before each trial so that the subject's next response can be recorded. The subject's blindfold is then removed, he is thanked for his cooperation, and cautioned to discuss the experiment with no one."

Illustration of the final steps of the planning (statistical treatment of the data, forming the evidence report, confronting the hypothesis with the evidence report, and generalization of the findings) can best be offered when these topics are later emphasized.

Writing Up an Experiment

After the experimenter has collected his data, subjected them to statistical analysis, and reached his conclusions, he writes up the experiment. The point of view taken here is that the same general format for writing up experiments should be used regardless of whether the report is to be submitted for publication in a scientific journal or whether it is a study conducted by a beginning class in experimental psychology. This maximizes the transfer of learning from a course in experimental psychology to the actual conduct of experiments as professional psychologists. The following is an outline that can be used for writing up the experiment. There are also offered a number of suggestions that should help to eliminate certain errors that students frequently make, and several other suggestions that should lead them to a closer approximation of scientific writing.

The main principle to follow in writing up an experiment is that the report must include every *relevant* aspect of the experiment—someone else should be able to repeat the experiment solely on the basis of the report. If this is impossible, the report is inadequate. On the other hand, the experimenter should not become excessively involved in details. Those aspects of an experiment which the experimenter judges to be irrelevant, should not be included in his report. In general, then, the report should include every important aspect of the experiment, but should also be as concise as possible—for scientific writing is economical writing.

The writer should also strive for clarity of expression. If an idea can be expressed simply and clearly, it should not be expressed complexly and ambiguously—"big" words or "high flown" phrases should be avoided wherever possible.

We shall adhere to certain standard conventions. The conventions and a number of additional matters about writing up an experiment may be found in the Publication Manual of the American Psychological Association. The close relationship between the write-up and the outline of the experimental plan should be noted. Frequent reference should be made to that outline in the following discussion for much of the write-up has already been accomplished there.

1. *Title.* The title should be short but indicative of the exact topic of the experiment. This does not mean that every topic included in the article should be specified in the title. The title needs to be unique—it should distinguish the experiment from all other investigations. Introductory phrases such as "A Study of . . ." and "An Investigation of . . ." should be avoided, since they are generally redundant.

2. *Name and Institutional Connection.* On the title page the author's name should be centered below the title, followed by his institutional connection. In the case of multiple authorship this sequence should be continued for all authors. If all authors are from the same institution, the affiliation should be listed last (and only once). In no case should the department within the institution be specified. Frequently an entire class conducts an experiment, in which case, strictly speaking, they are multiple authors. Since the main purpose of such class experiments is to provide practice for the individual student, however, it is suggested that only the name of the student writing up the experiment be used as the author, rather than listing the entire class including the professor.

3. *Introduction.* It was noted previously that the write-up of the literature survey could serve as the introductory section of the report. We said that the problem should be developed logically, citing all relevant studies. A summary statement of the problem should then be made, preferably as a question. Let us emphasize that the results of the litera-

ture survey should lead quite smoothly into the statement of the problem. For instance, if the experiment concerns the effects of alcohol on performance of a cancellation task (e.g., striking out all letter E's in a series of letters), you might cite the results of previous experiments which show detrimental effects of alcohol on various kinds of performance. At this point you might indicate that there is no previous work on the effects of alcohol on the cancellation task and that the purpose of your experiment was to extend the previous findings to that task. Accordingly the problem is, "Does the consumption of alcohol detrimentally affect performance on the cancellation task?" The hypothesis may also be developed logically, but it too should be summarized in one sentence, preferably in the "If . . . then . . ." form. It is not customary to label the introductory section; rather, it should simply start as the first part of the article.

4. *Method.* The main function of this section of the report is to tell your reader precisely how the experiment was conducted. Put another way, this section serves to specify the method of gathering data that are relevant to the hypothesis and that will serve to test the hypothesis. It is here that the main decisions need to be made as to which matters of procedure are relevant and which are irrelevant. If the author has specified every detail that is necessary for someone else to repeat the experiment, but no more, he has been successful. To illustrate, let us assume that a "rat" study has been conducted. The author would want to tell the reader that, say, a *T* maze was used, and then go on to specify the precise dimensions of the maze, the colors used to paint it, the type of doors, and the kind of covering. He would presumably not want to relate that the maze was constructed of pine, or that the wood used was one inch thick, for it is highly unlikely that these variables would influence the subject's performance. That is, it would be a strange phenomenon indeed if one could show that rats performed differently in a *T* maze depending on whether the maze (well painted) was constructed of pine, redwood, or walnut, or whether the walls were 3/4 inch or one inch thick.

The outline used in presenting the method is not rigid, and may be modified for each experiment. In general, however, the following information should be found, and usually in the following order.

a. *Subjects.* The population should be specified in detail, as well as the method of drawing the sample studied. If any subjects from the sample had to be "discarded" (students didn't show up for their appointments, they couldn't perform the experimental task, rats died, etc.) this information should be included—for the sample may not be random because of these factors.

b. *Apparatus.* All relevant aspects of the apparatus should be included.

Where a standard type of apparatus is used (e.g., a "Hull-type Memory Drum"), only its name need be stated. Otherwise, the apparatus has to be described in sufficient detail for another experimenter to obtain or construct it. It is good practice for the student to include a diagram of the apparatus in the write-up, although in professional journals this is only done where the apparatus is complex and novel.

c. *Design.* The type of design used should be included in a section after the apparatus has been described. The method of assigning subjects to groups, and the labels attached to the groups, are both indicated (experimental group, control group, etc.). The variables contained in the hypothesis need to be (operationally) defined; it is also desirable (at least for students) to indicate which are the independent and dependent variables. The techniques of exercising experimental control may be included. For example, if there was a particularly knotty variable that needed to be controlled, the techniques used for this purpose may be discussed.[8]

d. *Procedure.* The procedure for conducting the data collection phase of the experiment should be set down in detail. You must include instructions to the subjects (if they are human), the maintenance schedule and the way in which they were "adapted" to the experiment (if they are infra human animals), how the independent variable was administered, and how the dependent variable was recorded.

5. *Results.* The data relevant to the test of the hypothesis are presented here. These data are summarized as a precise sentence (the evidence report). If the data are in accord with the hypothesis, then it may be concluded that the hypothesis is confirmed. If they are not of the nature predicted by the hypothesis, then the hypothesis is disproved.[9]

It is usually advisable to present a *summary* of the data under "results."

[8] The inclusion of the definition of variables, experimental control, etc. in this section is arbitrary. You may well include separate sections for these matters, consider them under "Procedure," or arrange them otherwise.

[9] We have not yet reached certain important matters that need to be included in this section. This advance information is summarized below for future reference. You need not worry about what this information means if it is unfamiliar, for it will become clear later. The advantages of including an outline of all relevant information in one place would seem to outweigh the disadvantage of prematurely presenting this information.

You should state the null hypothesis as it applies to your experiment and also the significance level that you have adopted. Then include the results of the statistical test, and indicate the appropriate probability level. For example: "The null hypothesis was: There is no difference between the means of the experimental and control groups on the dependent variable. The t test yielded a value of 2.20. With 16 degrees of freedom, this value is significant beyond the 5 per cent level." (An alternative way of saying this would be: "The resulting t was 2.20 ($P. < .05$)," in which case we may assume that the degrees of freedom are obvious from the number of subjects used. However, for practice, it is a good idea for you to specify the number of degrees of freedom used. You may then continue: "It is therefore

This can almost always be done by using a table, but frequently figures can also be used to advantage. Whether or not tables and/or figures are used depends on the type of data and the ingenuity and motivation of the writer. Since students frequently are confused about the definitions of table and figure, as well as about their respective formats, we shall consider them in detail. Tables and figures are used for *summarizing* the data. They are not used for presenting *all* the data (so-called "raw data," a term that implies that the data have not been statistically treated). Nor are the steps in computing the statistical tests (the actual calculations) included under "results." In student write-ups, however, it has been found advisable to include the raw data and the steps in the computation of the statistical tests in a special appendix. The advantage of using such an appendix is that the instructor can correct any errors made in this part of the operation.

A table in the results section consists of numbers which summarize the main findings of the experiment. It should present these numbers systematically, precisely, and economically. A figure, on the other hand, is a graph, chart, photograph, or like material. It is particularly appropriate for certain kinds of data; for instance, to show the progress of learning.

In constructing a table, one should first determine what is to be shown —the main points that should be made apparent from the table. Then should be considered the question of what is the most economical way of making these points in a meaningful fashion. Since the main point of the experiment is to determine if certain relationships exist between certain variables, the table should show whether or not this relationship exists. Of course, it is possible to present more than one table, and tables may be used for purposes other than presenting data. For example, it is frequently possible to make the over-all design of the experiment more apparent by presenting the separate steps in tabular form (this use of tables is particularly recommended for students as it helps to "pull the experiment together" for them). To illustrate the format of a table, let us consider an experiment in which we study how smoking a cigarette affects steadiness. We shall say that as soon as the subjects finished their cigarettes, they took the Whipple steadiness test (see p. 53). Tables 4.1 and 4.2 might be used for this experiment.

Note that the previously stated requirements of a good table are clearly satisfied in these examples. Also observe the precise format used,

---

possible to reject the null hypothesis. Since this finding is in accord with the empirical hypothesis, we may conclude that that hypothesis is confirmed." Of course, if the empirical hypothesis predicted that the null hypothesis would be rejected, but it was not, then it may be concluded that the hypothesis was not confirmed.

Table 4.1

Outline of the Experimental Design

Phase	Experimental Group	Control Group
Independent Variable	Smokes one cigarette	Does not smoke
Dependent Variable	Takes Whipple Steadiness Test	Takes Whipple Steadiness Test

Table 4.2

Mean Steadiness Scores

	Experimental Group	Control Group
n*	12	14
Mean	22.20	15.27

* The symbol n stands for the number of subjects in each group. In this hypothetical experiment there were twelve subjects in the experimental group and fourteen in the control group.

for students have a habit of ignoring the details of the standardized conventions illustrated here.

The same general principles stated for the construction of tables holds for figures. In particular, a figure is used primarily to illustrate a relationship between the variables in the experiment. In showing the amount of steadiness exhibited on each of several successive trials Figure 4.1 (see p. 53) might be used.

Incidentally, we might reach certain tentative conclusions about the effect of our independent variable by a rough glance at Figure 4.1 (final conclusions would have to await the application of the appropriate tests of significance). We might note first that both curves decrease as the subjects have more trials, this suggesting that both groups become steadier as the number of trials increase. This is a reasonable conclusion, since we would expect the subjects to become steadier with practice and with adaptation to the experimental situation. Second, we note that the experimental subjects are less steady throughout the experiment than are the control subjects. Thus, it would seem reasonable to conclude tentatively that one effect of smoking is to decrease steadiness. Third, we might observe that the curve for the experimental subjects decreases more rapidly than the curve for the control subjects. This would lead to the conclusion that the effects of smoking are rather pronounced at first, but that they tend to wear off over a period of time. After ten trials, however, it appears that the experimental subjects are still less steady. But will the curves eventually come together? If we run the experiment again and give the subjects a larger number of test trials, we probably would find out. (Would twenty be enough?) The question of what would happen to the relative position of the curves if the subjects had been run on more trials seems to be a perennial one in experimentation. One might accept this as a lesson in planning his experiment: if he is going to be concerned about this question, take it into consideration before the data are collected. The unfortunate aspect of the question is that one seldom knows how many trials are enough—particularly before the experiment is run, and frequently even after the data are collected.

The above information should be of considerable help to you, but in case you would like further information about the construction of figures and graphs, a large number of good sources are available (e.g., Guilford, 1956).

6. *Discussion.* The main functions of this section are to interpret the results of the experiment and to relate these results to other studies. The interpretation involves essentially an attempt to explain the results. Perhaps some existing theory can be brought to bear to help understand the findings. If the hypothesis was derived from some general theory, then the confirmation of the hypothesis serves to strengthen that theory, and the findings in turn are explained by that hypothesis in conjunction with the larger theory. If the findings are contrary to the hypothesis, then some new explanation is required—to account for the results there may be advanced new hypotheses that run counter to the hypothesis tested. Or it may be that the hypothesis tested can be modified in such a way to make it consistent with the results. In this case a "patched up" hypothesis is advanced for future test.

In relating the results to other studies, the literature survey may again be brought to bear. By considering the present results along with previous ones new insights may be obtained—the results of this experiment may provide the one missing piece that allows the solution of the puzzle.

Hypotheses may also be advanced about any unusual deviation in the results. Referring to Figure 4.1, for instance, one may wonder why there was a sudden "bump" in both curves on trials 7 and 8. Is this a reliable "bump?" If it is, why did it occur? In short, what additional problems were uncovered that require further investigation?

There may be certain limitations in the experiment. If so, this is the place to discuss them, e.g., what variables apparently were inadequately controlled? How would one modify the experiment if it were to be repeated?

You may also consider the extent to which the results may be generalized. To what populations may you safely generalize? To what extent are the generalizations limited by uncontrolled variables, etc.?

A rather strange characteristic of some experimenters seems to be that they feel "guilty" or "embarrassed" when they have obtained negative results.[10] But whatever the reason, it is not appropriate to include long "alibis" for negative results. It is reasonable, however, to briefly speculate about why they were obtained.

7. *Summary.* The final section of the actual write-up should include a brief statement of the problem, the hypothesis, how the hypothesis was subjected to test, and the conclusions reached.

8. *References.* References to pertinent studies throughout the text should be made by citing the author's (or authors') last name, the year of publication and enclosing these in parentheses. If the name of the author occurs in the text cite only the year of publication in the parentheses. The references should then be listed alphabetically at the end of the paper. The form and order of items for journal references is as follows: last name, initials, title of the study, abbreviations for the name of the journal, year of publication of the study, volume number, and pages. For example, if two references like the following are used they might be referred to in the write-up as ". . . According to Lewis (1953), learning theory has already been shown to have applicability to human behavior. Examples of this position are numerous (e.g., Calvin, Perkins, and Hoffman, 1956) . . ." Then, in the "reference" section of the write-up, they would be *precisely* listed as:

References

Calvin, A. D., Perkins, M. J., and Hoffman, F. K. The effect of non-differential reward and non-reward on discriminative learning in children. *Child Develpm.*, 1956, 27, 439-446.
Lewis, D. J. "Rats and Men," *Amer. J. Sociol.*, 1953, 59, 131-135.

Some "Dos" and "Don'ts"

Finally, here are a few suggestions and bits of information that did not fit conveniently into the previous sections. These matters are directed at students in the hope of improving their reports. We could not pretend to be very inclusive here, but shall simply point out items that continue to appear every year with new students. If some of them seem minor, then it should be all the more easy for "erring" students to modify their behavior so that they more closely conform to conventional stand-

[10] "Negative results" occur when the hypothesis makes a particular prediction, but the results are contrary to that prediction. The term may also be used to indicate that the null hypothesis (p. 87) was not rejected, usually these two definitions amount to the same thing.

ards. This particular request for conformity is simply for the purpose of facilitating communication in scientific writing.

The first thing to do before writing up a report is to consult several psychological journals. Since we are concerned mainly with experimentation, you might start with the Journal of Experimental Psychology, although experiments are certainly reported in a large number of journals (these journals may be found in college libraries). Select several articles in these journals and study them rather thoroughly, particularly as to format. Try to note examples of the suggestions that we have offered, and particularly try to get the over-all idea of the continuity of the article. But be prepared for the fact that you will probably not be able to understand every point in each article. Even if there are large sections that you do not understand, do not worry too much about it for this understanding will come with further learning—by the time you finish this book you will be able to understand most professional articles. Of course, some articles are extremely difficult, or in fact impossible, to understand even by specialists in the field. This is usually so because of the poor quality of the write-ups.

In their own write-ups, students frequently make assertions such as: "*It is a proven fact* that left-handed people are steadier with their right hands than right-handed people are with their left hands," or "*Everyone knows that* sex education for children is good." Perhaps the main benefit to be derived by students in general from a course in experimental psychology is a healthy hostility for such statements. Before making such a statement you should have the data to back it up—you should cite a relevant reference. It is not wise to use such trite phrases as "It is a proven fact that . . ." or "Everyone knows that. . . ." The use of such phrases usually indicates a lack of data on the part of the user. If you want to express one of these ideas, but lack data, the ideas still can have a place in the introductory section. They simply should be stated more tentatively, e.g., "It is possible that left-handed people are steadier with their right hands than right-handed people are with their left hands," or "An interesting question to ask is whether left-handed people . . . etc." Our main point, then, is that if you want to assert that something is true, make sure that you have the data (a reference) to back it up— mere opinions asserted in a positive fashion are insufficient.

Another point about writing up reports is that personal references should be kept to a minimum.[11] For instance, students frequently say such things as "I believe that the results would have turned out differently, if I

[11] While the standard convention in scientific writing is to avoid personal references, there are those who hold that such a practice should not be sustained. The following quotation from an article entitled "Why are medical journals so dull?" states the case for this view: ". . . avoiding 'I' by impersonality and circumlocution leads to dullness and I would rather be thought conceited than dull. Articles are written to

had . . ." or "It is my opinion that . . ." Strictly speaking the scientific audience doesn't care too much about your emotional experiences, what you believe, feel, think, etc.—they are much more interested in what data you obtained and what conclusions you can draw from those data. Rather than stating what you believe, then, you should say something like "the data indicate that . . ." The report of an experiment falls within the context of justification rather than in the context of discovery.

Harsh or emotionally loaded phrases should also be avoided. The report of scientific work should be as divorced from emotional stimuli as possible. An example of bad writing would be: "Some psychologists believe that a few people have extrasensory perception, *while others claim it to be nonsense.*"

Misspelling occurs all too frequently in student reports. You should take the trouble to read the report over after it is written, and rewrite it if necessary. If you are not sure how to spell a word, look it up in the dictionary. Unfortunately, far too few students have acquired "the dictionary habit," a habit which is the mark of an "educated person."

When studying the format for writing up articles by referring to psychological journals, please note what sections are literally labeled. For instance, the introduction is not usually labeled as such, but the method section is always indicated, as are its sub-sections such as *Subjects, Procedure,* etc.

A few final matters are:

1. Don't list minor pieces of "apparatus," like a pencil (unless it is particularly important, as in a stylus maze).

2. Standard abbreviations that are used in the journals should be adhered to—S stands for "subject" and E stands for "experimenter" (note that they are capital letters).

3. The word "data" is plural. "Datum" is singular. Thus it is incorrect to say "that data," "this data," etc. Rather one should say: "those data," "these data," etc. Similarly "criterion" is singular and "criteria" is plural. Thus say—"This criterion may be substituted for *those* criteria."

4. There is a difference between a probability value (e.g., $P = .05$) and a percentage value (e.g., 5%). While a percentage can be changed into a probability and vice versa, one would not say that "The probability was 5%," or "The per cent was .05" if he really meant 5%.

5. When reporting the results of a statistical analysis, never say that "the data are (or are not) significant." Data are not significant in the technical sense. Rather, the results of your statistical analysis may be significant, e.g., a given value of t is significant.

---

interest the reader, not to make him admire the author. Overconscientious anonymity can be overdone, as in the article by two authors which had a footnote, 'Since this article was written, unfortunately one of us has died' " (Asher, 1958, p. 502).

6. If it is at all possible, the report should be typed, not written in longhand. Studies have shown that students who type, get higher grades. We shall not consider why this is so, except to point out that one possible reason is that instructors have a "better unconscious mental set" in reading typed papers.

7. When you quote from an article, put quotation marks around the quote and cite a page reference.

5

Experimental Design:
The Case of Two
Randomized Groups

A general understanding of how to conduct experiments should now be apparent. In Chapters 1 and 4 we attempted to cover all of the major phases of experimentation. In presenting an over-all picture of experimentation, however, it has been necessary to cover a number of steps hastily. The remaining chapters of the book will consist of attempts to fill in these relatively neglected areas. But we should try never to lose sight of how the steps on which we momentarily concentrate fit into the general picture of designing and conducting an experiment.

We shall now focus on the phase of experimentation that concerns the selection of a design. While there are a number of designs available to the experimenter, we have thus far limited our consideration to one that involves only two groups. This chapter shall discuss this type of design more thoroughly. Since the "two-groups" design is basic in psychology, an understanding of it will form a sound foundation from which we can move to more complex (though not necessarily more difficult to comprehend) designs.

A General Orientation

The "two-groups" design may more completely be referred to as the "two-randomized-groups-design." To summarize briefly what has been said about this design, let us recall that the experimenter defines an independent variable that he seeks to vary in at least two ways. (Incidentally, the two values that he assigns to the independent variable may be referred to as two "conditions," "treatments," or "methods"). He then seeks to determine whether these two conditions differentially affect his dependent variable. The general procedure that he follows to answer this question may be summarized as follows. First, he defines a certain population about which he wishes to make a statement. Then he randomly selects a sample of subjects to study. Since that sample has been drawn randomly from the population it may be assumed to be representative of the population—thus whatever is found to be true for the sample may be inferred to be true for the population. Let us assume that the population is defined as all students in a certain university. They may number six thousand. We decide that our sample shall be sixty in size. One reasonable method for selecting this sample would be to obtain an alphabetical list of the 6,000 students and take every one-hundredth student on that list. On the assumption that all sixty students will cooperate, we now have our sample. It has been specified that we will study two conditions in our experiment. To assign a separate group of subjects to each condition, the next problem is to divide the sixty subjects into two groups. Again, any method that would assure that the subjects are randomly assigned to the two groups would suffice. Let us say that we write the name of each subject on a separate slip of paper and place all sixty pieces of paper in a hat. We may then decide that the first name drawn would be assigned to the first group, the second to the second group, the third to the first group, etc. In this manner we would end up with two groups, each of thirty subjects. A simple flip of a coin would then tell us which is to be the experimental group, and which the control group. The reason that we referred to this type of design as the "two-randomized-groups-design" should now be apparent: subjects are *randomly* assigned to *two* groups. A better understanding of a randomized groups design will be possible when we contrast it with a type of design in which subjects are not randomly assigned to groups (see Chapter 8).

A basic and important assumption made in any type of design is that the means (averages) of the groups do not differ significantly at the start of the experiment. In a two-group design the two values of the independent variable are then respectively administered to the two groups. For example, some positive amount of the independent variable might be administered to the experimental group, while a zero amount is administered to the control group. Scores of all subjects on the dependent variable are then recorded and subjected to statistical analysis. If the appropriate statistical test indicates that the two groups are significantly different on the (dependent variable) scores, it may be concluded that this difference is due to the variation of the independent variable— assuming that the proper experimental controls have been in effect, it may be concluded that the two values of the independent variable are effective in producing the differences in the dependent variable.

"Equality" of Groups Through Randomization

Now, by randomly assigning the sixty subjects to two groups, we said, it is highly reasonable to assume that the two groups are essentially equal; but equal with respect to what? The answer might be that the groups as wholes are equal in all respects. And such an answer is easy to defend, assuming that the randomization has been properly carried out. In any given experiment, however, we are not interested in comparing the two groups in all respects. Rather, we want them to be equal on those factors that might affect our dependent variable. Suppose the dependent variable concerns the rate at which a person learns a task that involves visual abilities. In this case we would want the two groups to be equal at least with respect to intelligence and visual acuity. More particularly, we would want the means of intelligence and visual acuity scores to be essentially the same. For both of these factors are likely to influence scores on our dependent variable.

Students frequently criticize the randomized groups design by pointing out that "by chance" we could end up with two very unequal groups. It is possible, they say, that one group would be considerably more intelligent, on the average, than the other group, that is, that one group would have a higher mean intelligence score. While such an event is indeed possible, it is unlikely, particularly if a large number of subjects is used in both groups. For it can be demonstrated that the larger the number of subjects randomly assigned to the two groups, the greater will be their "equality" or "similarity." Hence, while with a small number of subjects it is unlikely that the two groups will differ to any great extent, it is more likely than if the number of subjects is large. The moral should be clear: if you wish to reduce the difference in the means of the two groups, use a large number of subjects.[1]

Even with a comparatively large number of subjects it is still possible, though unlikely, that the means of the groups will differ considerably by chance. Suppose, for example, that we have drawn a sample of sixteen subjects and assigned them to two groups. Now, if we measured their intelligence, it *is* possible that we would obtain a mean intelligence quotient of 100 for one group and a mean of 116 for the second group. However, by using appropriate statistical techniques we can determine that such an event should occur less than about five times out of one hundred, i.e., if we ran the experiment one hundred times, and assigned subjects to two groups at random in each experiment, a difference be-

---

[1] In making this point we are ignoring the distributions of the scores. While the matter is not as simple as we have made it, this presentation offers the main line of reasoning.

tween the groups of 16 IQ points (i.e., 116-100) or more should occur in only about five of the experiments. Differences between the two groups of less than sixteen IQ points should occur more frequently. And differences between the two groups of twenty-four points or more should occur less than one time in a hundred experiments, on the average. Most frequently, then, there should be only a small difference between the two groups.

"But," the skeptical student continues, "suppose that in the particular experiment that I am conducting (I don't care about the other ninety-five or ninety-nine experiments) I *do* by chance divide my subjects into two groups of widely differing ability. I would think that the group with the mean IQ of 116 would have higher mean scores on the dependent variable than does the other group, *regardless of the effect of the independent variable*. I (the experimenter) would then conclude that the independent variable is effective, when, in fact, it isn't."

One cannot help but be impressed by such a convincing attack, but retreat at this point would be premature, for there are still several weapons that can be brought into the battle. First, if one has doubts as to the "equalness" of his two groups, he can compute their scores on certain variables to see how their means actually compare. Thus, in the above example, we could actually measure the subjects' IQs and visual acuity, compute the means, for both groups, and compare the scores to see if there is much difference. If there is little difference, we know that our random assignment has been at least fairly successful. This laborious and generally unnecessary precaution actually has been taken in a number of experiments.[2]

But the student continues tenaciously. "Suppose I find that there is a sizeable difference, and furthermore, suppose that I determine this only after all data have been collected. My experiment would be invalidated." Yet, there is hope. In this case we could use a statistical technique that allows us to equate the two groups with respect to intelligence. That is, we could "correct" for the difference between the two groups and determine whether they differ on the dependent variable for a reason that cannot be attributed to the difference of intelligence. Put another way, we could statistically equate the two groups on intelligence so that differences on this extraneous variable would not differentially affect the dependent variable scores. Further consideration of this statistical technique (known as the analysis of covariance) would be premature here (see p. 291).

---

[2] In an experiment on rifle marksmanship, for instance, it was determined that four groups did not differ significantly on the following extraneous variables: previous firing experience, left or right handedness, visual acuity, intelligence, or educational level (McGuigan and MacCaslin, 1955, a).

"Excellent," the student persists, "but suppose the two groups differ in some respect for which we have no measure, and that this difference will sizeably influence scores on the dependent variable. I now understand that we can probably 'correct' for the difference between the two groups on factors such as intelligence and visual acuity, because these are easily measurable variables. But what if the groups differ on some factor which we cannot measure, or do not think to measure. In this case we would be totally unaware of the difference and illegitimate conclusions from our data."

"You," we say to the student, secretly admiring his demanding perseverance, "have now put us in such an unlikely position that we need not worry about its occurrence. Nevertheless, it *is* possible, just as it is possible to be hit by a car while crossing the street. And, if there is some factor for which we cannot make a 'correction,' the experiment will result in erroneous conclusions." The only point we can refer to here is one of the general features of the scientific enterprise: Science is self correcting. Thus, if any given experiment leads to a false conclusion, and if the conclusion has any importance at all for psychology, an inconsistency between the results of the invalid experiment and additional data will become apparent. The existence of this problem will then lead to a solution, which, in this case, will be a matter of discarding the incorrect conclusion.

Statistical Analysis of the Two-Randomized-Groups Design[3]

Now, the matter for us to discuss here concerns a return to a question that was briefly considered in Chapter 1, where we posed the following problem: After the experimenter has collected his data on the dependent variable, he wishes to determine whether one group is superior to the other. His hypothesis may predict that the experimental group will be superior to the control group. The first step in testing the hypothesis might be to compute the mean scores on the dependent variable for the two groups. It might be found that the experimental group has a higher mean score than the control group—say that the experimental group has a mean score of forty, while the control group has one of thirty-five. Assuming that the higher the score the better the performance, can we conclude that this five-point difference is significant? Or is it merely a chance difference? To answer this question, we said, we must apply a statistical test. Let us now consider one statistical test that is frequently used to answer this question.

The statistical test to which we refer is known as the "t test" (note

[3] Before beginning this section the conscientious student might want to read the first section of Chapter 14, "Concerning the Accuracy of the Data Analysis."

that a lower-case "t" is used to denote this test, not a capital "T," which has another denotation in statistics).[4] The first step in computing a t-test value is the computation of the means of the dependent variable scores of the two groups concerned. The equation for computing a mean (symbolized $\overline{X}$) is

$$(5.1) \qquad \overline{X} = \Sigma X/n$$

The only unusual symbol in equation 5.1 is Σ, the capital Greek letter sigma. Σ may be interpreted as "sum of." It is a summation sign and simply instructs you to add whatever is to the right of it.[5] In this case the letter X is to the right of sigma so we must now find out what values X stands for and add them. Here, X merely indicates the score that we obtained for each subject. Suppose, for instance, that we gave a test to a class of five students, with these resulting scores:

	X
Joan	100
Constance	100
Richard	80
Lillian	70
Joe	60

To compute ΣX we merely need to add up the X scores. In this way we find that $\Sigma X = 410$ ($100 + 100 + 80 + 70 + 60 = 410 = \Sigma X$). The n in equation 5.1 stands for the number of subjects in the group. In this example, then, $n = 5$. Thus, to compute $\overline{X}$ we simply substitute 410 for ΣX, 5 for n in equation 5.1, and then divide n into ΣX:

$$\overline{X} = 410/5 = 82.00$$

Thus the mean score of the group of 5 students who took the particular test is 82.00.

Let us now turn to an equation for computing t:[6]

[4] Before our first discussion of the statistical analysis of an experimental design, it is well to point out that the statistical tests (such as the t test) are conducted on the assumption that certain statistical assumptions are satisfied. Since the assumptions for all the statistical tests discussed in this book are similar, it is more economical to discuss them together after all of our designs have been considered (p. 285). The instructor or student who so wishes, of course, may immediately integrate this topic with the discussion of the statistical tests.

[5] More precisely, Σ instructs you to add all the values of the symbols that are to its right, values that were obtained from your sample.

[6] This equation is applicable when the number of subjects in the two groups does not differ appreciably. For a completely general equation to use when this assumption cannot be satisfied, and for another example of the computation of t, see p. 93. Also for a related discussion where the variances of the two groups are not equal, see Li (1957, p. 91). It might be added that the author conducted an experiment using three different equations. Students performed significantly better (in terms of errors and time to solve for t) using equation 5.2 than they did using the others.

$$(5.2) \qquad t = \frac{\overline{X}_1 - \overline{X}_2}{\sqrt{\dfrac{(\Sigma X_1^2/n_1) - (\overline{X}_1)^2}{n_1 - 1} + \dfrac{(\Sigma X_2^2/n_2) - (\overline{X}_2)^2}{n_2 - 1}}}$$

While this formula may look forbidding to the statistically naive, such an impression should be short-lived for t is actually rather simple to compute. To illustrate, assume that we are testing a hypothesis to the effect that "psychology students are more anxious than English students." After following all the rules for collecting data, assume that we have the following scores (the higher the score, the higher the anxiety):

Psychology Students (Group 1)		English Students (Group 2)	
Subject Number	Anxiety Score	Subject Number	Anxiety Score
1	10	1	5
2	10	2	4
3	8	3	3
4	7	4	3
5	7	5	1
	$\Sigma X_1 = 42$		$\Sigma X_2 = 16$

We now seek to obtain an evidence report, i.e., a summary statement of the findings of the study. This evidence report, then, will tell us whether the hypothesis is probably true or false. The first step is to compute the means of the two groups. We denote the psychology students as Group 1, and the English students as Group 2. Note that subscripts have been used in equation 5.2 to indicate which group the various values are for. In this case ΣX_1 stands for the sum of the scores for the first group, and ΣX_2 for the sum of the scores for the second group. In like manner $\overline{X}_1$ and $\overline{X}_2$ are the means for groups 1 and 2 respectively, and n_1 and n_2 are the respective number of subjects in the two groups. We can now determine that $\Sigma X_1 = 42$, while $\Sigma X_2 = 16$. Since the number of subjects in Group 1 is five, we note that $n_1 = 5$. The mean for Group 1 (i.e., $\overline{X}_1$) may now be determined by substitution in equation 5.1:

$$\overline{X}_1 = 42/5 = 8.4$$

And similarly for Group 2 (n_2 also is 5):

$$\overline{X}_2 = 16/5 = 3.2$$

We now need to compute ΣX_1^2 and ΣX_2^2 for equation 5.2.

These two symbols indicate that we should sum the squares of each score for each group respectively. Thus, to compute ΣX_1^2 we should square the score for the first subject, add it to the square of the score for the second subject, add both of these values to the square of the score for the third subject, etc. Squaring the scores for the subjects in both groups and summing them we obtain:

Psychology Students			English Students		
Subject Number	X_1	X_1^2	Subject Number	X_2	X_2^2
1	10	100	1	5	25
2	10	100	2	4	16
3	8	64	3	3	9
4	7	49	4	3	9
5	7	49	5	1	1
		362			60

One frequent error by students should be pointed out as a precaution. That is that ΣX^2 is *not* the square of ΣX. That is, $(\Sigma X)^2$ is not equal to ΣX^2. For instance, the $\Sigma X_1 = 42$. The square of this value is $(\Sigma X_1)^2 = 1764$, whereas $\Sigma X_1^2 = 362$. We may now summarize the various values required by equation 5.2 as:

$$\overline{X}_1 = 8.4 \qquad \overline{X}_2 = 3.2$$
$$n_1 = 5 \qquad n_2 = 5$$
$$\Sigma X_1^2 = 362 \qquad \Sigma X_2^2 = 60$$

And substituting these values in equation 5.2 we obtain:

$$t = \frac{8.4 - 3.2}{\sqrt{\dfrac{(362/5) - (8.4)^2}{5 - 1} + \dfrac{(60/5) - (3.2)^2}{5 - 1}}}$$

We now need to go through the following steps in computing t:

1. Obtain the differences between the means: $8.4 - 3.2 = 5.2$
2. Divide ΣX_1^2 by n_1: $362/5 = 72.40$
3. Square $\overline{X}_1$: $8.4 \times 8.4 = 70.56$
4. Divide ΣX_2^2 by n_2: $60/5 = 12.00$
5. Square $\overline{X}_2$: $3.2 \times 3.2 = 10.24$
6. Compute $n_1 - 1$: $5 - 1 = 4$
7. Compute $n_2 - 1$: $5 - 1 = 4$

The results of these operations are:

$$t = \frac{5.2}{\sqrt{(72.40 - 70.56)/4 + (12.00 - 10.24)/4}}$$

In the next stage subtract as indicated:

$$t = \frac{5.2}{\sqrt{1.84/4 + 1.76/4}}$$

and divide:

$$t = \frac{5.2}{\sqrt{0.46 + 0.44}}$$

then sum the values in the denominator:

$$t = \frac{5.2}{\sqrt{0.90}}$$

The next step is to find the square root of 0.90. This may be obtained from page 298 in the Appendix, and is found to be 0.95. Dividing as indicated we find t to be:

$$t = 5.2/.95 = 5.47$$

Although the computation of t is straightforward, the beginning student is likely to make an error in its computation. The error is generally not one of failing to follow the procedure, but one of a computational nature (dividing incorrectly, failing to square terms properly, mistakes in addition, etc.). A great deal of care must be taken in statistical work; each step of the computation should be checked in an effort to eliminate errors. As an aid to the student in learning to compute t, a number of exercises are provided at the end of this chapter. Work all of these exercises and make sure that your answers are correct.

One point in the computation of t needs to be clarified. In the numerator we have indicated that $\overline{X}_2$ should be subtracted from $\overline{X}_1$. Actually we are conducting what is known as a "two-tailed test." You need not be concerned about this term here, but the important point for you to observe is that we are interested in the absolute difference between the means. Hence $\overline{X}_1 - \overline{X}_2$ is appropriate if $\overline{X}_1$ is greater than $\overline{X}_2$ ($\overline{X}_1 > \overline{X}_2$). But if in your experiment you find that $\overline{X}_2$ is greater than $\overline{X}_1$ ($\overline{X}_2 > \overline{X}_1$) then you merely subtract $\overline{X}_1$ from $\overline{X}_2$, i.e. equation 5.2 would have as its numerator $\overline{X}_2 - \overline{X}_1$.

We might also note that the value under the square root sign is always positive. If it is negative in your computation, go through your work to find the error.

The reason we want to obtain a value of t, we said, is to determine whether the difference between the means of two groups is a chance difference or whether it is a significant difference. But several additional matters must also be discussed in relation to this difference. The first is a consideration of what is known as the "null hypothesis," a concept that it is vital to understand. The null hypothesis that is generally used in psychological experimentation, roughly, states that there is no difference between two groups. Since we wish to contrast the two groups by comparing their means on the dependent variable, we may more precisely state that there is no difference between the population means on the dependent variable of the two groups.

The null hypothesis, let it be emphasized, states that there is no difference between the *population* means.[7] The reason we conduct an experiment is to make statements about populations—to determine whether the population means of our two groups differ. In a sense this may be stated otherwise as follows: we want to know whether the *true* means of our groups differ (where the true mean is taken as the population mean). Now, of course, we cannot study the population directly. Rather, the only way to determine whether or not the true (population) means differ is by comparing the means obtained for our two sample groups. We do this by subtracting one sample mean from the other, as specified in the numerator of equation 5.2. This difference will almost certainly not be zero; but it will be some positive amount, the value of which may be quite small or quite large. If the difference between our sample means is quite small, we would be inclined to conclude that the difference is due to chance. On the other hand, if the difference is large, then we might say that the difference is too large to be due to chance. Thus the null hypothesis asserts that the difference between the population means is zero. In effect it says that any difference between the means of the groups in your sample is due to chance. If we find that the difference between the means of our groups is small, then it follows that that difference is a chance difference, and that the null hypothesis is reasonable. But if our groups differ considerably, then the difference is probably too large to be due to chance, and the null hypothesis is not tenable in that particular case.

The question now is how small the difference must be between $\overline{X}_1$ and $\overline{X}_2$ before we can say that it is a chance difference. Then, too, how large must the difference be before we can say that it is *not* a chance difference? The latter question can be answered by the value of t; if t is sufficiently large, we can say that the difference between the two groups is too large to be attributed to chance. And to determine how large "sufficiently large" is we may consult the table of t. But before doing this, there is one additional value that we must compute—the degrees of freedom (df). The degrees of freedom available for this application of the t test are a function of the number of subjects in the experiment. More specifically, $df = N - 2$.[8] And N is the number of subjects in one group (n_1) plus the number of subjects in the other group (n_2). Hence, in our example we have:

[7] A symbolic statement of the null hypothesis would be $\mu_1 - \mu_2 = 0$ (μ is the Greek letter mu). Here μ_1 is the population mean for group 1 and μ_2 is the population mean for group 2. If the difference between the sample means ($\overline{X}_1 - \overline{X}_2$) is small, then we are likely to infer that there is no difference between the population means; thus, that $\mu_1 - \mu_2 = 0$. On the other hand if $\overline{X}_1 - \overline{X}_2$ is large, then the null hypothesis that $\mu_1 - \mu_2 = 0$ is probably not true.

[8] This equation for computing df is only for this application of the t-test. We shall use other equations for df when considering additional statistical tests.

$$N = n_1 + n_2 \quad \text{i.e.,} \quad N = 5 + 5 = 10$$

hence: $\qquad df = 10 - 2 = 8$

To determine whether our t is significant, let us now turn to a table of t (Table 5.1) armed with two values: $t = 5.47$ and $df = 8$. The table of t is organized around two values: a column labeled "df" and a row labeled "P" (for probability). The df column is on the extreme left, and the P row runs across the top of the table. Values of t take up the rest of the columns (rows). Our general purpose here is to find out what P value is associated with a specific value of t and df. To do this we must first run down the df column until we arrive at our specific value of df; in this case, 8 df. We then read across the row that is marked 8 df. This row contains a number of possible values of t—0.130, 0.262, 0.399, etc. We must read across this row until we come to a value of t that is close to our particular value—in this case, 5.47. But the largest value of t in this row is 3.355; this value, then, is the closest match we can make to 5.47. So, we read up the column that contains 3.355 to determine what value of P is associated with it—in this case, .01.

Let us make a general observation: the larger the t, the smaller the P. For example, with 8 df a t of 0.130 has a P of 0.9 associated with it, while with the same df a t of 1.860 has a P of 0.1. From this observation and our study of the tabled values of t and P we can conclude that if a t of 3.355 has a P of 0.01, any t larger than 3.355 must have a smaller P than .01. It is sufficient for our purposes simply to note this fact without attempting to make it any more precise.

The next step is to interpret the fact that a t of 5.65 has a P of less than 0.01 ($P < 0.01$) associated with it. This finding indicates that a difference between means of the two groups of the size that we obtained has a probability of less than 0.01; i.e., that a difference between the means of this size may be expected less than one time in a hundred by chance. Put another way, if we conducted the experiment one hundred times, by chance we would expect a difference of this size to occur once, provided the null hypothesis is true. This, we must all agree, is a most unlikely occurrence. It is so unreasonable, in fact, to think that such a large difference could have occurred by chance on the very first of our hypothetical one hundred experiments that we prefer to reject "chance" as an explanation. And since our null hypothesis is our expression of the "chance explanation," it is unreasonable to regard this as tenable. We therefore choose to reject our null hypothesis. That is, we refuse to regard it as reasonable that the real difference between our two groups is zero when we have obtained such a large actual difference, as indicated by the respective means, in this case, of 8.4 and 3.2. But if a difference of this size cannot be attributed to chance, what reason can we give for it? We assume that all the proper safeguards of experimentation have been

Table 5.1*

Table of t

df	P = 0.9	0.8	0.7	0.6	0.5	0.4	0.3	0.2	0.1	0.05	0.02	0.01
1	0.158	0.325	0.510	0.727	1.000	1.376	1.963	3.078	6.314	12.706	31.821	63.657
2	0.142	0.289	0.445	0.617	0.816	1.061	1.386	1.886	2.920	4.303	6.965	9.925
3	0.137	0.277	0.424	0.584	0.765	0.978	1.250	1.638	2.353	3.182	4.541	5.841
4	0.134	0.271	0.414	0.569	0.741	0.941	1.190	1.533	2.132	2.776	3.747	4.604
5	0.132	0.267	0.408	0.559	0.727	0.920	1.156	1.476	2.015	2.571	3.365	4.032
6	0.131	0.265	0.404	0.553	0.718	0.906	1.134	1.440	1.943	2.447	3.143	3.707
7	0.130	0.263	0.402	0.549	0.711	0.896	1.119	1.415	1.895	2.365	2.998	3.499
8	0.130	0.262	0.399	0.546	0.706	0.889	1.108	1.397	1.860	2.306	2.896	3.355
9	0.129	0.261	0.398	0.543	0.703	0.883	1.100	1.383	1.833	2.262	2.821	3.250
10	0.129	0.260	0.397	0.542	0.700	0.879	1.093	1.372	1.812	2.228	2.764	3.169
11	0.129	0.260	0.396	0.540	0.697	0.876	1.088	1.363	1.796	2.201	2.718	3.106
12	0.128	0.259	0.395	0.539	0.695	0.873	1.083	1.356	1.782	2.179	2.681	3.055
13	0.128	0.259	0.394	0.538	0.694	0.870	1.079	1.350	1.771	2.160	2.650	3.012
14	0.128	0.258	0.393	0.537	0.692	0.868	1.076	1.345	1.761	2.145	2.624	2.977
15	0.128	0.258	0.393	0.536	0.691	0.866	1.074	1.341	1.753	2.131	2.602	2.947
16	0.128	0.258	0.392	0.535	0.690	0.865	1.071	1.337	1.746	2.120	2.583	2.921
17	0.128	0.257	0.392	0.534	0.689	0.863	1.069	1.333	1.740	2.110	2.567	2.898
18	0.127	0.257	0.392	0.534	0.688	0.862	1.067	1.330	1.734	2.101	2.552	2.878
19	0.127	0.257	0.391	0.533	0.688	0.861	1.066	1.328	1.729	2.093	2.539	2.861
20	0.127	0.257	0.391	0.533	0.687	0.860	1.064	1.325	1.725	2.086	2.528	2.845
21	0.127	0.257	0.391	0.532	0.686	0.859	1.063	1.323	1.721	2.080	2.518	2.831
22	0.127	0.256	0.390	0.532	0.686	0.858	1.061	1.321	1.717	2.074	2.508	2.819
23	0.127	0.256	0.390	0.532	0.685	0.858	1.060	1.319	1.714	2.069	2.500	2.807
24	0.127	0.256	0.390	0.531	0.685	0.857	1.059	1.318	1.711	2.064	2.492	2.797
25	0.127	0.256	0.390	0.531	0.684	0.856	1.058	1.316	1.708	2.060	2.485	2.787
26	0.127	0.256	0.390	0.531	0.684	0.856	1.058	1.315	1.706	2.056	2.479	2.779
27	0.127	0.256	0.389	0.531	0.684	0.855	1.057	1.314	1.703	2.052	2.473	2.771
28	0.127	0.256	0.389	0.530	0.683	0.855	1.056	1.313	1.701	2.048	2.467	2.763
29	0.127	0.256	0.389	0.530	0.683	0.854	1.056	1.311	1.699	2.045	2.462	2.756
30	0.127	0.256	0.389	0.530	0.683	0.854	1.055	1.310	1.697	2.042	2.457	2.750
	0.12566	0.25335	0.38532	0.52440	0.67449	0.84162	1.03643	1.28155	1.64485	1.95996	2.32634	2.57582

* Table 5.1 is reprinted from Table IV of Fisher: *Statistical Methods for Research Workers*, published by Oliver and Boyd Ltd., Edinburgh, by permission of the author and publishers.

observed in obtaining these results, and that the groups therefore differed only in the respect that each was administered a different amount of this variable. It seems reasonable to assert, then, that the reason the two groups differed is that they received different amounts of the independent variable. This leads to the further conclusion that the independent variable is effective in influencing scores on the dependent variable; and this is precisely what we sought to determine by conducting the experiment.

There are still a number of questions about this procedure that need to be answered. For instance, we said that before we conduct an experiment it is unlikely that we would (by chance) obtain a P of .01 for our t. How small may P be before it can be considered sufficiently likely to occur by chance? That is, how small must P be before we can reject the null hypothesis? For example, with 8 df, if we had obtained a value of 1.90 for t, we find that the corresponding value of P is less than .10. This means that a difference between the two groups could be expected by chance less than ten times out of one hundred. Now, is this sufficiently unlikely that we can reject the null hypothesis? Again, consider a t of 2.31. The corresponding P is less than 0.05. Can we reject the null hypothesis on the basis of this size of P? What if we had obtained a t of 0.90, with a corresponding P of less than 0.40; a difference of this size may be expected forty times out of one hundred. Is this P sufficiently small to allow us to reject the null hypothesis? In short, the question is this: what value of P is small enough to allow us to reject the null hypothesis? Unfortunately, there is no simple answer to this question, for it depends on a number of things. The best we can say here is that the experimenter may set any value of P as the cut-off point. Thus he may say that: "If the value of t that I obtain has a P of less than 0.50, I will reject my null hypothesis." Similarly, he may set P at 0.01, 0.05, 0.30, or even 0.99 if he wishes. There is only one requirement that he must satisfy in setting the value of P: he must set it *before* he conducts his experiment. The reason for this is that, for a proper test of the null hypothesis, the experimenter should not be influenced by the particular nature of his data. For example, it would be inappropriate to run a t test and determine P to be 0.06, and then decide that if P is 0.06 to reject the null hypothesis. Such an experimenter would be inclined never to fail to reject the null hypothesis, for the criterion (the value of P) for rejecting it would be determined by the value of P actually obtained. An extreme case of this would be a person who obtains a P of 0.90, and sets 0.90 as his criterion. The sterility of such a decision is apparent, for a difference between his groups of the size that he obtained would be expected by chance ninety times out of one hundred. Obviously, it would be unreasonable to reject the null hypothesis with such a large P, and such an experimenter would almost surely be committing an error, i.e., he would be rejecting the null hypothesis when in fact it should not be rejected.

The P that an experimenter sets, then, is totally arbitrary. He can vary it with the particular experiment that he is conducting. For some problems it is important to have an extremely small P, while for others a larger one is appropriate. But while the actual decision is arbitrary, there are a number of important considerations that will help the experimenter in arriving at his decision. The interested student will find such matters discussed in elementary statistics courses. Suffice it to say here that for general psychological experimentation a standard value of P is accepted —0.05. Thus, unless the experimenter specifies otherwise, he generally sets a P of .05 before conducting his experiment.[9]

Let us now apply the above considerations to our example. Our hypothesis was to the effect that psychology students are more anxious than English students. It was found that the mean scores for the two groups were 8.4 and 3.2, respectively. Furthermore, we found that the t-test yielded a value of 5.47, which, with 8 df, had a P of less than 0.01. Assuming the conventional value of 0.05 for P as our criterion of whether or not to reject the null hypothesis, our value of less than .01 causes us to reject the null hypothesis. That is, we assert that there is a true difference between our two groups. Furthermore, we observe that the direction of the difference is that specified by our (empirical) hypothesis, i.e., our hypothesis predicted that the scores for psychology students would be higher than for English students. Since the scores are of the nature predicted by the hypothesis (and significantly so), we may conclude that the hypothesis is confirmed.

The following general rule may now be stated: *If the empirical hypothesis predicts that there will be a difference between two groups, and if the null hypothesis is rejected, and if the difference between the two groups is in the direction specified by the hypothesis, then it may be concluded that the empirical hypothesis is confirmed.* Thus, there are two cases in which the hypothesis would not be confirmed: first, if the null hypothesis were not rejected; and second, if it were rejected, but if the difference between the two groups were in the opposite direction specified by the hypothesis. To illustrate these latter possibilities, let us assume that we actually obtained a t of 1.40 (which you can see has a P of less than 0.20). We fail to reject the null hypothesis, and accordingly fail to confirm the empirical hypothesis. On the other hand, assume that we obtain a t of 2.40 ($P < 0.05$), but that the mean score for the English students is higher than that for the psychology students. In this case we reject the null hypothesis, but fail to confirm the empirical hypothesis.

Strictly speaking, equation 5.2 is appropriate for computing t only when the number of subjects in each of the two groups is equal (i.e.,

[9] The value of P set as the criterion of rejecting or failing to reject the null hypothesis is known as the *level of significance*. A similar term, *confidence level,* should not be substituted for it since the latter has a different meaning in statistics.

when $n_1 = n_2$). When the number of subjects in each of the two groups differ somewhat, however, the error in t is negligible and may be ignored. We have presented this rather generally applicable equation (equation 5.2) because of its greater ease. In an experiment in which the number of subjects in the two groups differ considerably, it would be wise to use a different equation.[10] Such a case would be, for instance, eighty subjects in one group and only twelve in the other. For further practice in computing t where the numbers of subjects in two groups are not equal but still do not differ greatly, consider the following problem.

Suppose that we are interested in the effects of a certain tranquilizing drug on psychoses. This drug is widely used but it is important to evaluate experimentally its effects: does it actually "help" psychotic patients? To answer this question we might design an experiment so that twelve subjects (experimental group) receive the tranquilizing drug, while twenty subjects (control group) do not. Assume that after an appropriate period of treatment we obtain the following scores, where the higher the score, the greater the psychotic tendency.

Scores of Psychotic Tendencies

Experimental Group		Control Group	
Subject	Score	Subject	Score
1	0	1	8
2	6	2	9
3	4	3	12
4	8	4	11
5	4	5	12
6	2	6	9
7	7	7	8
8	2	8	2
9	3	9	1
10	6	10	9
11	9	11	11
12	8	12	12
		13	9
		14	10
		15	8
		16	6
		17	7
		18	3
		19	9
		20	10

[10] The following is probably the simplest equation to use in such a case. You should be able to compute t from it on the basis of the information presented above.

$$t = \sqrt{\frac{(N-2)(n_2\Sigma X_1 - n_1\Sigma X_2)^2}{N[n_1n_2(\Sigma X_1^2 + \Sigma X_2^2) - n_2(\Sigma X_1)^2 - n_1(\Sigma X_2)^2]}}$$

Where N is the total number of subjects in both groups; i.e., $N = n_1 + n_2$.

We shall use equation 5.2 for computing our t, since the n's of the two groups do not differ appreciably.

Computing the necessary values for equation 5.2 we find:

	Experimental Group[11]		*Control Group*
$\overline{X}_E =$	4.92	$\overline{X}_C =$	8.30
$n_E =$	12	$n_C =$	20
$\Sigma X_E =$	59	$\Sigma X_C =$	166
$\Sigma X_E^2 =$	379	$\Sigma X_C^2 =$	1570

And substituting these values in formula 5.2:

$$t = \frac{8.30 - 4.92}{\sqrt{\dfrac{(379/12) - (4.92)^2}{12 - 1} + \dfrac{(1570/20) - (8.30)^2}{20 - 1}}}$$

Performing the indicated operations

$$t = \frac{3.38}{\sqrt{(31.58 - 24.21)/11 + (78.50 - 68.89)/19}}$$

$$= 3.38/1.09 = 3.10$$

We thus find that $t = 3.10$. With thirty degrees of freedom ($N - 2 = 32 - 2 = 30$) we refer to our table of t, and find that the P associated with it is less than 0.01. Since this P is less than our conventional criterion of 0.05, we reject the null hypothesis and conclude that there is a difference between the groups. On the assumption that the experiment was properly conducted we may conclude that the administration of the tranquilizing drug led to this difference. Since the control group had a higher mean score than the experimental group, we may further conclude that the administration of the tranquilizer led to a reduction in psychotic tendencies.

At this point it may be beneficial to discuss further the null hypothesis. This hypothesis is a statistical hypothesis and is set up for the purpose of attempting to disprove it. Our null hypothesis asserts that there is no difference between the population means of our two groups; we seek to determine that it is false, that there is a difference between the means. Hence, if it is disproven in a properly conducted experiment, we can conclude that there is a difference between our two groups and furthermore that this difference is due to variation of the independent variable. If we cannot disprove the null hypothesis, then we cannot assert that there is a difference between the two groups; variation of our independent variable is not effective.

Incidentally, two characteristics of the null hypothesis may be il-

---

[11] The subscripts identify the two groups, e.g. $\overline{X}_E$ is the mean for the experimental group, ΣX_C is the sum of the scores for the control group.

lustrated by reference to the word *null*. First, it may be noted that "null" derives from the Latin *nullus*, "not any." Hence, a purpose of experimentation is to determine whether the difference between the experimental result and that specified by the null hypothesis is null. If there is a "null" discrepancy between the experimental results and the null hypothesis, then we do not reject the null hypothesis. If, however, the discrepancy is positive, is not "null," then we reject it.

The second association is with the word "nullify." In this case the null hypothesis is the one that we seek to nullify. If we can nullify (reject) it, then we may conclude that the independent variable is effective. But if we fail to nullify it, the opposite is true.[12]

Let us now summarize each major step that we have gone through in testing the empirical hypothesis.

1. State the hypothesis, e.g., "If the anxiety scores of English and psychology students are measured, the psychology students will have the higher scores."

2. The experiment is designed according to the procedures outlined in Chapter 4, e.g., "anxiety is operationally defined [such as scores on the Manifest Anxiety Scale, Taylor, 1953], samples from each population are drawn, etc.

3. The null hypothesis is stated—"there is no difference between the mean scores of the two groups."

4. A probability value for determining whether or not to reject the null hypothesis is established, e.g., if $P < .05$, then the null hypothesis will be rejected; if $P > .05$, the null hypothesis will not be rejected.

5. The data are collected and statistically analyzed. For this design a t test is conducted whereby the means of the two groups are determined. The value of t is computed and the corresponding P ascertained.

6. If the means are in the direction specified by the hypothesis (if the psychology students have a higher mean score than the English students)

---

[12] The term "null hypothesis" was first used by Professor Sir Ronald A. Fisher (personal communication). He chose the term 'null hypothesis' without "particular regard for its etymological justification but by analogy with a usage, formerly and perhaps still current among physicists, of speaking of a null experiment, or a null method of measurement, to refer to a case in which a proposed value is inserted experimentally in the apparatus and the value is corrected, adjusted, and finally verified, when the correct value has been found; because the set-up is such, as in the Wheatstone Bridge, that a very sensitive galvanometer shows no deflection when exactly the right value has been inserted.

"The governing consideration physically is that an instrument made for direct measurement is usually much less sensitive than one which can be made to kick one way or the other according to whether too large or too small a value has been inserted.

"Without reference to the history of this usage on physics. . . . One may put it by saying that if the hypothesis is exactly true no amount of experimentation will easily give a significant discrepancy, or, that the discrepancy is null apart from errors of random sampling."

and if the null hypothesis is rejected, it may be concluded that the hypothesis is confirmed. If the null hypothesis is not rejected, it may be concluded that the hypothesis is not confirmed. Or, if the null hypothesis is rejected, but the means are in the direction opposite to that predicted by the hypothesis, then the hypothesis is not confirmed. (Hence, in our example the empirical hypothesis is confirmed.)

"Borderline" Significance

One frequently occurring problem in experimentation is that of borderline significance. An experimenter who sets a P of 0.05 as his criterion for rejecting the null hypothesis fails to reject the null hypothesis if he obtains a P of 0.30. But suppose that he obtains a P of 0.06. One might argue that, "Well, this isn't quite .05 but it is so close that I'm going to reject the null hypothesis anyway. This seems reasonable; after all, this means that a difference between groups of the size that I obtained can be expected only six times out of a hundred by chance. Surely this is not much different than a probability of five times out of a hundred." To this there is only one answer: the t test is *decisive*—a P of 0.06 is *not* a P of 0.05. In this case, therefore, there is no alternative but to fail to reject the null hypothesis. If the experimenter had set up a criterion of a P of 0.06 *before* he conducted his experiment, then we would have no quarrel with him—he could, in this event, reject his null hypothesis. But since he established a criterion of a P of 0.05, he cannot modify his criterion after the data are collected, not even if he obtains a P of 0.051.

At the same time, however, we must agree with this experimenter that a P of 0.06 is an unlikely event by chance. Our advice to him is: "Yes. It looks like you *might* have something. It's a good hint for further experimentation. Conduct a new experiment and see what happens. If, in this replication you come out with significant results, you are quite safe in rejecting the null hypothesis. But if the value of t obtained is quite far from significant in this new, independent test, then you have saved yourself from making an error."

The Method of Systematic Observation

We have previously contrasted two types of investigations: "experiments" and "systematic observation studies." The alert reader probably noticed that the first example used to illustrate the computation of t was a systematic observation study, while the second was an experiment. For in the study concerning anxiety of students of psychology and English *no variable was produced and purposively manipulated by the investi-*

gator. Rather, the study concerned observations of a phenomenon that was *already present in the population*. To meet the requirements for an experiment in that example we would have had to assign subjects randomly to two groups and then decree that everyone in one group would major in psychology and all those in the other, in English. If we had been able to do this, we could say that our independent variable was "major of the student" and that we had varied it in two ways: English and psychology majors. Since we did not vary it in this manner, it was not purposively manipulated, and hence the study cannot be said to be an experiment. In the second example, however, the requirements of experimentation *were* fulfilled, since we randomly assigned subjects to two groups and then determined which group would receive the tranquilizing drugs. Hence, the independent variable was the amount of tranquilizing drugs administered, varied in two ways: (1) a zero amount and (2) some positive amount. Since the independent variable was under the control of the experimenter and since he *induced* different conditions in the two groups, it may be said that he *purposively manipulated* the independent variable and thus conducted an experiment.

Judgment of the importance of the difference between the two types of investigations should be held in abeyance until we consider "control" (see Chapter 6). It will be shown that there can be important differences in the two types of investigations as far as confidence in their respective conclusions is concerned. The main point to observe here is that the *t* test may be an appropriate method of statistical analysis for both types of investigations. And this point should be fairly apparent from our two examples.

Summary of the Computation of t for a Two-Randomized-Groups Design

Assume that we have obtained the following dependent variable scores for the two groups of subjects:

Group 1	Group 2
10	8
11	9
11	12
12	12
15	12
16	13
16	14
17	15
	16
	17

1. We start with equation 5.2, the equation for computing t:

$$t = \frac{\overline{X}_1 - \overline{X}_2}{\sqrt{\frac{(\Sigma X_1^2/n_1) - (\overline{X}_1)^2}{n_1 - 1} + \frac{(\Sigma X_2^2/n_2) - (\overline{X}_2)^2}{n_2 - 1}}}$$

2. Compute the sum of $X(\Sigma X)$, the sum of X^2 (ΣX^2), and n for each group.

	Group 1	Group 2
ΣX	108	128
ΣX^2	1512	1712
n	8	10

3. Using equation 5.1, compute the means for each group.

$$\overline{X}_1 = 108/8 = 13.50 \qquad \overline{X}_2 = 128/10 = 12.80$$

4. Substitute the above values in equation 5.2.

$$t = \frac{13.50 - 12.80}{\sqrt{\frac{(1512/8) - (13.50)^2}{8 - 1} + \frac{(1712/10) - (12.80)^2}{10 - 1}}}$$

5. Perform the operations as indicated and determine the value of t.

$$t = 0.70/\sqrt{0.96 + 0.82} = 0.53$$

6. Determine the number of degrees of freedom associated with the above value of t.

$$df = N - 2 = 18 - 2 = 16$$

7. Enter the table of t, and determine the probability associated with this value of t. In this example $0.70 > P > 0.60$. Therefore, assuming a significance level of 0.05, the null hypothesis is not rejected.

Problems

1. An experimenter runs a well-designed experiment wherein $n_1 = 16$ and $n_2 = 12$. He obtains a t of 2.14. Assuming that he has set a significance level of 0.05, can he reject his null hypothesis?

2. An experimenter obtains a computed t of 2.20 with 30 df. The means of his two groups are in the direction indicated by his empirical hypothesis. Assuming that the experiment was well designed and that the experimenter has set a significance level of 0.05, did his independent variable influence his dependent variable?

3. It is advertised that a certain tranquilizer has a curative effect on psychotics. A clinical psychologist seeks to determine whether or not

this is the case. He conducts a well-designed experiment and obtains the following results on a measure of psychotic tendencies. Assuming that he has set a significance level of 0.01, and assuming that the lower the score the greater the psychotic tendency, determine whether the tranquilizer has the advertised effect.

Scores of group that received the tranquilizer	Scores of group that did not receive the tranquilizer
2, 3, 5, 7, 7, 8, 8, 8	1, 1, 1, 2, 2, 3, 3

4. A psychologist hypothesizes that people who are of similar body build work better together. Accordingly, he forms two groups. Group 1 is composed of individuals who are all tall and lean, while Group 2 consists of individuals who are short and stout. He has both groups perform a task that requires a high degree of cooperation. The performance of each subject is measured where the higher the score, the better the performance on the task. He sets a significance level of 0.02. Did he confirm or disconfirm his empirical hypothesis?

Group 1	*Group 2*
10, 12, 13, 13, 15, 15, 15, 17, 18, 22, 24, 25, 25, 25, 27, 28, 30, 30	8, 9, 9, 11, 15, 16, 16 16, 19, 20, 21, 25, 25, 26, 28, 29, 30, 30, 32, 33, 33

5. On the basis of his experience, a marriage counselor suspects that when one spouse is from the North and the other is from the South the marriage has a likelihood of being unsuccessful. He selects two groups of subjects: Group 1 composed of marriage partners both of whom are from the same section of the country (either North or South), and Group 2 consisting of marriage partners from the North and the South respectively. He sets a 0.05 level of significance and obtains ratings of the success of the marriage (the higher the rating, the better the marriage). Assume that adequate controls have been effected. Is his suspicion confirmed?

Group 1	*Group 2*
1, 1, 1, 2, 2, 3, 3, 4, 4, 5, 6, 6, 7, 7, 7	1, 1, 2, 3, 4, 4, 5, 5, 6, 7

6

Experimental Control

The Nature of Experimental Control

Experimental control is one of the most important phases in the planning and conduct of experiments and therefore requires particular vigilance on the part of the experimenter. The word "control" indicates that the experimenter has a certain power over the conditions of his experiment. We may generally think of experimental control as the manipulation of the variables involved in the experiment. In order to clarify the meaning of control, imagine that the visitor from Mars descends to Earth for the limited purpose of determining whether a feather, when all support is withdrawn, will fall to the ground. Being from Mars, this man has no knowledge about terrestrial events. To accomplish his task he climbs to the top of a building and releases the feather. He then observes the feather as it floats through the air, eventually going out of sight on its way into the clouds.

He returns to Mars with his report that feathers behave peculiarly on Earth—when released, they go up. At this point his Martian brothers may ask a number of questions concerning the methods that he used in arriving at this conclusion: Is there air on Earth? If so, might not air movements have caused the feather to climb? Were any sounds present? Maybe on Earth sounds have a peculiar effect on such objects as feathers. What about the presence of lights? Might they not have affected the motion of the feather? And so forth. To these questions our Martian friend can only shrug his shoulders in embarrassment. It is obviously possible that the behavior of the feather was effected by such extraneous variables. If this is so, then the attracting force of the ground might have been cancelled by their power. If such extraneous variables had not been allowed to influence the feather, it might have actually fallen to the ground.

It is the judgment of the Martian authorities that the experiment is inconclusive. Being a persistent soul, however, our Martian friend obtains a new grant from the Foundation for the Advancement of Knowledge of the Earth (commonly referred to as FAKE by government offi-cials) and descends once more to Earth, this time with considerably more foresight. On arrival he sets about constructing a rather tall apparatus, air tight so that he can withdraw all of the air from it, leaving a vacuum inside. Furthermore, it is soundproof, lightproof, resistant to radioactivity, etc. The Martian then climbs to the top, releases the feather, and observes that it falls to the ground. Returning to Mars he reports his results to the Martian Academy of Science where his colleagues are much more re-ceptive than before.

Let us now consider these two experiments with reference to the prob-lem of control. First, our friend exercised control over his independent variable in both cases. That is, he produced the event that he wished to study. This is the first sense in which we shall use the word "control." We shall say that an experimenter exercises *independent-variable-control* when he varies the independent variable in a known and specified man-ner. In our examples he has determined whether or not to drop the feather in the presence of the ground. (Recall that independent-variable-control is the essential defining feature of an *experiment,* as distinguished from a *systematic observation.*)

The second sense in which we shall use "control" may be made ap-parent by restating the purpose of the experiments: the Martian sought to determine whether the position of the feather (when released) was related to the position of the ground. He was interested in determining whether these two variables were related. There were also present, how-ever, a number of other (extraneous) variables that might have affected the position of the feather. If there *is* a relationship of the type he sought, it might have been hidden from him by these other variables. From our knowledge of conditions on earth we might suppose that movements of the air constituted the important extraneous variable that prevented our friend from success in his first experiment. Even without our knowledge his Martian colleagues were able to suggest other possible distorting effects—light, sound, radioactivity, etc. For with no knowledge of the effects of these extraneous variables it was necessary to assume that they might affect the position of the feather. The second experiment was more successful, for in that experiment the effect of these extraneous variables was eliminated. And once their effects were removed, the Martian ob-tained the relationship that he sought. In the first experiment, then, no effort was made to regulate the extraneous variables while in the second experiment they *were* regulated. Hence, we shall say that our second sense of "control" refers to the regulation of extraneous variables, and shall call this kind of control *extraneous-variable-control.*

In order to be clear we shall say that an extraneous variable is one that is operating in the experimental situation in addition to the independent variable. Since it might affect the dependent variable, and since we are not immediately interested in it, it must be regulated so that it will not mask the possible effect of the independent variable. Furthermore, we shall say that if an extraneous variable is allowed to operate in the experimental situation in an uncontrolled manner it and the independent variable are *confounded* (the dependent variable is not free from irrelevant influences). And if confounding is present, then the reason that any change occurs in the dependent variable cannot be ascribed to the independent variable. Hence, in the first experiment we may say that the independent variable was confounded with air movements, with light, with sound, etc.

To illustrate further these two senses of "control," and in particular to "get closer to home," consider a psychological example. Suppose that an experimenter is interested in determining the effect of Vitamin A on certain visual abilities. The independent variable might be operationally defined as the amount of Vitamin A administered according to a certain schedule. The dependent variable might be similarly defined as the number of letters that a subject can see on a chart placed some distance from him. Since the independent variable is under the control of the experimenter he may vary it as he wishes. He may, for instance, vary it in three ways: one group of subjects may receive a placebo but no Vitamin A; a second group may receive a total of three ounces of the Vitamin, while a third group is administered a total of five ounces (over a period of time). In this way he is exercising control of the independent variable.

To illustrate extraneous-variable control we might note that the lighting conditions under which the test is taken are relevant to the number of letters that the subjects can correctly report. Suppose, for example, that the vision test is taken in a room in which the amount of light varies considerably during the day, and further that Group 1 is run mainly in the morning, Group 2 around noon, and Group 3 in the afternoon. In this case some subjects would take the test when there is good light, others when it is poor. The test scores might then primarily reflect the lighting conditions rather than the amount of Vitamin A administered, in which case the possible effects of Vitamin A would be masked out. Put another way, the amount of lighting and amount of Vitamin A would be confounded. Lack of control over this extraneous variable would leave us in a situation where we do not know which variable or combination of variables is responsible for influencing our dependent variable.

Just to develop this point briefly, let us consider some of the possibilities when only the single extraneous variable of light is uncontrolled. Assume that the obtained value of the dependent variable increases as the amount of Vitamin A increases, i.e., that the group receiving the five-ounce dose of Vitamin A has the highest dependent variable score, the three-ounce group is next, and that the zero Vitamin A group has the lowest test score. What may we conclude about the effect of Vitamin A on the dependent variable? Since light is uncontrolled we do not know what effect it has. Hence, the light may actually be the factor that causes the dependent variable scores to increase. Or, it is possible that lighting has a detrimental effect such that if it were not operating in an uncontrolled fashion the apparent effects of Vitamin A would be even more pronounced, e.g., if the five-ounce group received a score of 10, it might have received a score of 20, if light had been controlled. Another possibility is that the light has no effect, in which case our results could be accepted as valid. But since we do not know this, we cannot reach such a conclusion. The ambiguity in interpreting the effects of an independent variable where a single extraneous variable is not controlled should thus be apparent. But where there is more than one extraneous variable that is uncontrolled, the situation is much nearer total chaos.

Experimental control, then, is the regulation of experimental variables. And we may consider two classes of experimental variables: independent and extraneous. The independent variables, we have said, are those whose effects the experimenter is attempting to determine—he wants to know if a given independent variable affects his dependent variable. The extraneous variables are all other variables operating on the subjects at the time of the experiment. By exercising independent-variable control the experimenter varies the independent variable as he wishes. By exercising extraneous-variable control he regulates the extraneous variables so that confounding is eliminated. If adequate extraneous-variable control is exercised, an unambiguous statement on the relationship between the independent and dependent variables can be made. If extraneous-variable control is inadequate, however, the conclusion must be tempered. The extent to which it must be tempered depends on a number of factors, but, generally, inadequate extraneous-variable control leads to no conclusion whatsoever concerning the relationship.

Determining Extraneous Variables

We know that at any given moment a fantastically large number of stimuli are impinging on an organism. And we must assume that all of these stimuli are affecting the organism's behavior. But in any given experiment we are usually interested in only one aspect of behavior—a single class of responses. Furthermore, we usually seek to determine whether a certain class of stimuli affect that response—this is the independent-dependent variable relationship. Hence, for this immediate purpose we want to eliminate from consideration all other variables. If this were possible we could conclude that any change in our dependent variable is due only to the variation of our independent variable. If these other (extraneous) variables are allowed to influence our dependent variable, however, any change in our dependent variable could not be ascribed to variation of our independent variable—*we would not know which of the numerous variables caused the change.*

We must, then, control the experimental situation so that these other, extraneous variables can be dismissed from further consideration. The first step in this process is to identify them: what extraneous variables may be present in the experimental situation? Since it would be an almost endless task to list all of the variables that might conceivably affect the behavior of an organism, our question must be more limited: of all the variables present, which might conceivably affect our dependent variable? While this is still a difficult question, we can immediately eliminate from consideration a large number of influences on the organism. For example, if we are studying a learning process, we would not even consider such variables as color of the chair in which the subject sits, brand of pencil he uses, etc. As a first step in determining those extraneous variables that should be considered, we might refer to our literature survey. We can study previous experiments concerned with our dependent variable to find out which variables have been demonstrated to affect that dependent variable. We should also note what other variables previous experimenters have considered it necessary to control. Discussion sections of earlier articles may also yield information about variables that had not previously been controlled, but were recommended for consideration in the future. Together with the results of our literature survey, our general knowledge of potentially relevant variables, and considerable reflection concerning other variables, we may arrive at a list of extraneous variables that should be considered.

Specifying Extraneous Variables to Be Controlled

Once our list of potentially relevant extraneous variables is constructed, we must decide which should be controlled. This would include those variables that are likely to affect our dependent variable. It is to these variables that the various techniques of control will be applied. A discussion of these techniques is presented on pp. 106-116. Suffice it to state, here, the end result—the changes in the dependent variable will be ascribable to the independent variable rather than to the controlled extraneous variables.

Specifying Extraneous Variables That Cannot Reasonably Be Controlled

A simple answer to the question of which extraneous variables should be controlled is that we should control all of them. However, while it *might* be possible to control all of them, such a feat would be too expensive in terms of time, effort, and money. For example, suppose that the variation in temperature during experimental sessions is five degrees. While it is possible to control this variable, it is highly unlikely (in most experiments) that it would significantly (if at all) affect the dependent variable. And the experimenter's effort would be great enough to make the game "not worth the candle." This is particularly so when one considers the large number of other variables in the same category. With the limited amount of energy and resources available to him, the experimenter should seek to control only those variables that he considers potentially relevant.

Now, all of these probably minor variables might accumulate to have a rather major effect on the dependent variable, thus invalidating the experiment. And even if the effect is not so extreme, should even a minor extraneous variable be allowed to influence the dependent variable? If the experimenter is not going to control them, what can he do about them? In thinking about these points, we must remember that there always will be a large number of variables in this category. The question is, will they affect one of our experimental conditions (one of our groups) to a greater extent than another? For, if by chance such variables do not differentially affect our groups, then our worries are considerably lessened. We can make the assumption that such variables will "randomize out," that, in the long run, they will affect both groups equally. If it is reasonable to make this assumption, then this type of variable should not delay us further. A further discussion of randomization as a technique of control will allow us to consider this and similar problems later.

When to Abandon the Experiment

Up to this point we have been rather optimistic. We have assumed that we are capable of controlling all of the relevant variables that affect the dependent variable, that the effects of these variables will be essentially equal on all groups in the experiment. If it is unreasonable to make this assumption, then the experimenter must ask himself if these variables are of sufficient importance to necessitate the abandonment of the experiment. Even if he is not sure on this point, perhaps it would be best if the experiment were not conducted. Sometime after considering the various control problems, the experimenter must ask himself what he will gain by conducting the experiment. In these cases of inadequate extraneous variable control, the answer need not be that nothing will be gained; for instance, by conducting such an experiment it may be that further insight or beneficial information will be acquired concerning the control problem. But the experimenter must realize that this situation exists and be realistic in understanding that it may be better to abandon the experiment.

Techniques of Control

We have previously emphasized the importance of exercising adequate experimental control, but this phase of the experimental planning is sufficiently important that it does not seem possible to *over*emphasize it. Although experimenters try to exercise considerable vigilance in this regard, it is frequently the case that a crucial, uncontrolled extraneous variable is discovered only after the data are compiled. Shortcomings in control are found even in published experiments. Certainly, if such variables have been discovered neither by the experimenter nor by the editors of the journals, they are elusive and subtle. Furthermore, errors of control are not the sole property of young experimenters; they may be found in the work of some of the most respected and established psychologists.

The experimenter should give as much thought to potential errors as possible. After he has checked and rechecked himself it may be possible to obtain critiques from colleagues. An "outsider" can sometimes approach the experiment with a totally different set, thus seeing something that the experimenter himself might have missed.

Our main consideration in this section follows from the point at which an important extraneous variable is spotted and the experimenter must ask himself how it is to be controlled. He must ascertain what techniques are available for regulating it in such a manner that the effects of the independent variable can be clearly isolated. There are a number of such techniques; we shall attempt to classify them into several categories. This classification will necessarily be incomplete and overlapping in part, particularly as to the variations of each class. But a general understanding of the major principles should facilitate their application to a wide variety of specific control problems.

1. *Elimination.* The most desirable way to control extraneous variables is to eliminate them from the experimental situation. The Martian, for example, had to deal with such extraneous variables as air movements, sound, light, radioactivity. These variables were allowed to operate in his first experiment but they were controlled by elimination in the second

experiment: He designed a container that eliminated all the variables but his feather and the ground.

This technique of control is used rather widely in psychological laboratories—as in sound-deadened rooms for human subjects and the Skinner box, which is sound-deadened and opaque. Unfortunately, the method of elimination is frequently inapplicable. The previous example concerning Vitamin A and the subject's ability to read letters from a chart is a case in point. In that example our extraneous variable was the amount of lighting. Obviously, if the method of elimination were applied, the subjects would not have the light needed for them to see the chart. Other extraneous variables that cannot be removed are subjects' previous experience, sex, level of motivation, age, weight, intelligence, and so on.

2. *Constancy of Conditions.* When certain extraneous variables cannot be eliminated, we can attempt to hold them constant throughout the experiment. Control by this technique means essentially that whatever the extraneous variable, the same value of it is present for all subjects. Perhaps, for instance, the time of day is an important variable. Maybe people perform better on the dependent variable early in the morning than late in the afternoon. In order to hold time of day constant we might introduce all subjects into the experimental situation at approximately the same hour on successive days. Of course this procedure would not really hold the amount of fatigue constant for all subjects on all days. But it would certainly help.

Another example of effecting constancy of conditions would be our Vitamin A–chart-reading experiment. In this case we would attempt to hold the lighting conditions constant. Thus, we might pull down the blinds in our experimental room and have the same light turned on for all subjects. In experiments where light intensity is extremely important we could actually measure the amount of light present for each subject. The placing of a rheostat in the lighting circuit would allow us to modify fluctuations in the electrical flow in such a manner as to hold light intensity at almost precisely the same value for all subjects. Or we might prefer to use a DC source of electricity for our light as it would not fluctuate.

One of the standard applications of the technique of holding conditions constant is to conduct experimental sessions in the same room. Thus whatever might be the influence of the particular characteristics of the room (gayness, odors, color of the walls and furniture, location), that influence would be the same for all subjects. In like manner to hold various subject variables constant (educational level, sex, age), we need merely select subjects with the characteristics that we want. For example we might specify that all subjects must have completed the eighth grade and no more, that they are all male, that they are all 50 years old, etc.

Numerous characteristics of our experimental procedure must be subjected to this technique of control. Instructions to subjects, for instance, are extremely important. For this reason experimenters read precisely the same written set of instructions to all subjects (except where they must be modified for different experimental conditions). But even if the same words are read to all subjects, they might be read in different ways, with different intonations and emphases, regardless of the experimenter's efforts to avoid such differences. To exercise more precise control, then, many experimenters have all subjects listen to the same standardized instructions from a tape recorder.

Procedurally, all subjects should go through the same steps in the same order. For instance, if the steps are: greet subject, seat him, read instructions, blindfold him, tell him to start, and so on—then one would not want to blindfold some subjects *before* the instructions were read and blindfold others *after* the instructions. The attitude of the experimenter should also be held as constant as possible for all subjects. If he is jovial with one subject, and gruff with another, confounding of experimenter attitude with the independent variable would occur. Now, acting the same toward all subjects is extremely difficult, but a strong effort should be made in this direction. The experimenter can practice the experimental procedure a number of times until it becomes so routine that he can treat each subject in a mechanical fashion. Of course, the same experimenter should collect data from all the subjects. If different experimenters are used unsystematically, then a rather serious error may result. This seems to be particularly important in the case of animal research. In one experiment, for instance, an experimenter ran a group of rats for fourteen days, but had to be absent on the fifteenth day. The rats' performance for a different experimenter on that day was sufficiently different from other groups who had not suffered a change of experimenters that it is reasonable to conclude that the mere handling of them by a new person (who undoubtedly used somewhat different methods of picking them up, etc.) was responsible for the change.

The apparatus that is used both in administering the experimental treatment and in recording the results should be the same for all subjects. Suppose, for example, that two memory drums are used in an experiment, one of which moves more slowly than the other. If some subjects use the fast drum and others the slower, confounding will result. Application of the technique of constancy of conditions dictates that all subjects use the same drum. Similar precautions should be taken with regard to recording apparatus.

3. *Balancing.* Sometimes it is not convenient or even possible to hold constant conditions in the experiment. The experimenter may resort to the technique of balancing out the effect of the extraneous variable.

There are two general situations in which balancing may be used: (1) where the experimenter is either unable or uninterested in identifying the extraneous variables; (2) where he can identify them and desires to take special steps to control them.

Consider the first situation. One group of experimenters were interested in the effect of rifle training on rifle steadiness; whether a prolonged period of training in rifle firing increased the steadiness with which soldiers held their weapons (McGuigan & MacCaslin, 1955b). Previous research had indicated that the steadier a man held a rifle, the more accurately he could shoot. Thus, if you could increase steadiness through rifle training, you *might* thereby increase rifle accuracy. The design of the experiment was a test of rifle steadiness before and after they received their rifle training. If the soldiers were steadier on the second test, it might be concluded that training increases steadiness. The first group of data that were analyzed are presented in Table 6.1, where the lower the score, the greater the steadiness.

Table 6.1

Mean Steadiness Scores of Soldiers Before and After Rifle Training

Before Training		After Training
235.39	Training Period	194.26

From Table 6.1 we can see that the scores actually did decrease. The first thought is to conclude that training increases steadiness. When the experimenters analyzed another set of data from a group who did not receive rifle training, the picture changed (see Table 6.2).

Table 6.2

Mean Steadiness Scores of Trained and Untrained Groups

	Before Training		After Training
Trained (Experimental) Groups	235.39	Training period	194.26
Untrained (Control) Group	226.61	No training period	170.33

From Table 6.2 we can see that not only did the steadiness scores of the untrained group also decrease, but that they decreased more than those of the trained group. In order to reach the conclusion that training is the variable responsible for the decrease in scores, the experimental group had to show a significantly greater decrease in scores than did the control (no training) group. Thus, we may say that rifle training was not the reason that the steadiness scores of the trained group decreased. There must have been other variables operating to produce that change, variables that operated on both the experimental and the control groups. Whatever the variables, they were controlled by the technique of balancing (i.e., their effects on the trained group were balanced out or equalized by the use of the control group). But we may speculate about these extraneous variables. For example, the rifle training was given during the first two weeks of the soldiers' army life. It may be that the drop in scores merely reflected a general adjustment to the emotional impacts of army life. Or the soldiers could have learned enough about the steadiness test in the first session, to improve their performance in the second.

But whatever the extraneous variables, they were controlled by the use of the control group. The logic of using a control group should now be apparent. If the experimental and control groups are treated in the same way except with regard to the independent variable, then any difference between the two groups on the dependent variable is ascribable to the independent variable (at least in the long run). Thus, we need not specify all the extraneous variables that influence the two groups during the experiment. For instance, suppose that only three extraneous variables operate on the experimental group in addition to some positive amount of the independent variable. By administering a zero amount of the independent variable to the control group and by balancing out the effects of the three extraneous variables by allowing them to operate also on the control group, we can see from Figure 6.1 that the independent variable is the only one that differentially influences the two groups.

An additional important use of the control group as a technique of control may now be profitably discussed. Granting that (1) a large number of extraneous variables are operating on a subject in any given situation and (2) we cannot remove all of these variables, *then* we can use additional control groups to evaluate the influence of these variables, to analyze the total situation into its parts. Referring to Figure 6.1 we may be interested in the effect of extraneous variable 1. To evaluate that extraneous variable we need only add an additional control group

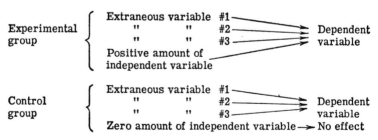

Figure 6.1. Diagramatic representation of the use of the control group as a technique of balancing.

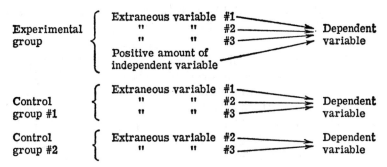

Figure 6.2. The use of a second control group to evaluate the effect of an extraneous variable.

which is not influenced by it (i.e., receives a zero value of it). The plan is illustrated in Figure 6.2.

For both the experimental group and control group 1 extraneous variable 1 is confounded with the other variables that are operating in the situation. Since this variable is not operating for control group 2, a comparison of the two control groups should tell us the effect of extraneous variable 1. Consider one of the extraneous variables that was operating in the rifle steadiness experiment: practice in the test situation. If an acquaintance with testing procedure and the specific learning of how to hold the rifle while tested, led to lower scores, the addition of control group 2 should provide us with this information. (See Table 6.3.)

Table 6.3

Possible Experimental Design for Studying the Effect
of Practice on the First Steadiness Test

	Receive Test Before Training?	Receive Training?	Receive Test After Training?
Trained (Experimental) Group	yes	yes	yes
Untrained (Control) Group 1	yes	no	yes
Untrained (Control) Group 2	no	no	yes

This is the same design as in Table 6.2 except for the addition of a second control group which does not take the initial test. A comparison of the steadiness scores of the two untrained groups on the test after training should tell us what the effect of the initial test is. If, for instance, untrained group 2 scores higher on the second (post-training) test than untrained group 1, we could say that the effect of taking the first (pre-training) test is to increase steadiness.

The use of additional control groups to evaluate the effect of various extraneous variables is a very important experimental technique. Frequently, the independent variable itself is of such a nature that it can be further fractionated by the addition of control groups. Some additional examples of this technique are taken up in the last section of this chapter.

The second situation in which balancing may be used is where there is a specific and known extraneous variable to be controlled. For instance, if an experimenter wishes to control the sex variable, he would use subjects of only one sex. If he has both male and female subjects available, however, and not enough subjects of one sex to use them exclusively, he may be forced to use both sexes. In this event he may control the effect of sex on his dependent variable by making sure that it balances out in his two groups. This would be accomplished by assigning an equal number of subjects of each sex to each group. If he has forty males and thirty females, he would make sure that twenty males and fifteen females go in each group.[1] Thus, if sex is relevant to his dependent variable, its effects would be the same for each experimental condition. In a similar manner he could control the age of the subjects—he would make sure than an equal number of each age group is assigned to each group.

The same holds true for apparatus problems. Suppose that an experimenter has two memory drums available and wants to use both to save time. They may or may not have slight differences, but to make sure that this variable is controlled he would have half of the subjects in each group use each memory drum. If he has thirty subjects in each of two groups, fifteen subjects in each group would use drum A, while the other fifteen would use drum B. Thus, whatever the differences in the memory drums, if any, their respective effects would be the same for both groups.

Balancing may also be applied where there is more than one experimenter. In this case we merely need to have each experimenter run an equal number of the subjects in each group. To consider a situation that is a bit more complicated than we have previously discussed let us say that we wish to balance two effects: sex and experimenter. We have two groups: sixty subjects per group, two sexes, and two experimenters. In this case the balancing arrangement would look something like that presented in Figure 6.3.

In a final example of the balancing technique, let us say that we want to see how well rats retain a certain habit. To do this we might use a T maze which has a white and a black goal box. From the start position, the rats may run to either box. But we wish to train them to run to one box, say the black. Hence, we always feed them when they run to the

---

[1] Of course, in this example, even though the effect of sex is balanced out, the males will have a greater effect on the dependent variable scores than the females. If this is disturbing to you in an experiment, the matter can be handled with appropriate statistical procedures.

Group I	Group II
15 Males - Experimenter #1	15 Males - Experimenter #1
15 Males - Experimenter #2	15 Males - Experimenter #2
15 Females - Experimenter #1	15 Females - Experimenter #1
15 Females - Experimenter #2	15 Females - Experimenter #2

Figure 6.3. Illustration of a design where experimenters and sex are balanced.

black box; they receive no food at the white box. After a number of trials they run rather consistently to the black box, ignoring the white box. After this initial training, we do not allow them to run the maze for, say, three months, at the end of which we place them in the start box of the maze for their test trials. If most of their runs are to the black box, may we assume that they "remembered" well? Our conclusion should not be so hasty. For we know that rats are nocturnal animals and that they tend to prefer dark places over light places. In particular, they have a preference for black goal boxes over white ones *before any training.* Hence it is possible that they would go more frequently to the black box on the test trials *regardless of the previous training.* For this reason we need to exercise control over the color of the "reward" box—we need to balance the colors. To do this we train half of our animals as above. The other half are trained in the opposite manner—they receive food when they run to the *white* box but no food when they run to the black box. If, on the test trials, the animals trained to go to white show a preference for white on the test trials then we would have considerably more confidence in our conclusion. For regardless of the color that they were trained to, we find that they retain the habit over our three month period—the effect of color cannot possibly be the variable that is influencing their behavior.

4. *Counterbalancing.* Some experiments are designed so that the same subject must serve under two or more different experimental conditions. If an experimenter were interested in whether a stop sign should be painted yellow or red, his problem would be to determine to which colored sign a subject responds faster. To answer this question he might measure a subject's reaction time to, first the red sign, and then the yellow sign. By repeating this procedure with a number of subjects he could reach a conclusion, perhaps that reaction time to the yellow sign is the smaller. Since the subjects were first exposed to the red sign, however, their reaction time to that sign would be partially dependent on their learning to operate the experimental apparatus and on their adaptation to the experimental situation. After they have learned how to operate the apparatus and adapted to the situation, they are exposed to the yellow sign. Hence, their lower reaction time to the yellow

might merely reflect practice and adaptation effects rather than effect of color—color of sign and amount of practice are confounded. The answer frequently given to the problem of how to control the extraneous variable of amount of practice is to use the method of counterbalancing. The application of this method would be to have half the subjects react to the yellow sign first and the red sign second, while the other half would experience the red sign first and the yellow sign second. (See Table 6.4.)

Table 6.4

Demonstrating Counterbalancing to Control an Extraneous Variable.
Experimental Session

	1	2
½ Subjects	Yellow Sign	Red Sign
½ Subjects	Red Sign	Yellow Sign

The general principle of the technique of counterbalancing may be stated as: Each condition (e.g., color of sign) must be presented to each subject an equal number of times, and each condition must occur an equal number of times at each practice session. Furthermore, each condition must precede and follow all other conditions an equal number of times. It can be seen that this principle is applicable to any number of conditions. For example, the principle of counterbalancing could be applied where we have three colors of signs.

If a number of experimental sessions are involved, not only would a certain amount of improvement in the subjects' performance due to practice be expected, but also a certain decrement in performance due to fatigue. The method of counterbalancing attempts to distribute these effects equally to all conditions. Hence, whatever the practice and fatigue effects, they presumably influence each condition equally since each condition occurs equally often at each stage of practice.[2]

We shall not consider the method further here because its application might entail difficulties. First, the practice effects from the red to the yellow sign might not be the same as those from the yellow sign to the red. For example, seeing the red sign first might induce a greater practice (or fatigue) effect, possibly leading to erroneous conclusions. Furthermore, the statistical analysis of experiments using this technique is rather troublesome, for the assumption of no interaction between conditions is difficult to meet (cf. Gaito, 1958), (see p. 211). Probably because of these difficulties, counterbalancing is relatively rare in the experimental literature. A design which does not expose the same subject to more than one condition is generally preferred, such as occurs with a

[2] For a more extended discussion of counterbalancing you are referred to Underwood (1949).

randomized-groups design. It would be a simple matter to have half the subjects react only to the yellow sign and the other half only to the red sign. Nevertheless, it is well to be aware of the technique for application to certain problems where the necessary assumptions can be satisfied.

To eliminate confusion let us specify the differences between balancing and counterbalancing. Balancing is used when each subject is exposed to only one experimental condition. Since the extraneous variable occurs equally often for each condition, the total effect of the extraneous variables is *balanced* or equalized for all conditions. Counterbalancing is used only when there are repeated measures on the same subject (although not in every case of repeated measures).

5. *Randomization*.[3] This technique is used for two general situations: (1) where it is known that certain extraneous variables operate in the experimental situation, but it is not feasible to apply one of the above techniques of control; (2) where we assume that some extraneous variables will operate, but cannot anticipate them and therefore cannot apply the other techniques.[4] In either case we take precautions that enhance the likelihood of our assumption that the extraneous variables will "randomize out," i.e., that whatever their effects, they will influence both groups to approximately the same extent.

Consider some examples. Subjects' characteristics are important in any psychological experiment. Such variables as previous learning experiences, level of motivation, amount of food eaten on the experimental day, relations with boy or girl friends, and money problems, may affect our dependent variable. Of course, the experimenter cannot control such variables by any of the previous techniques. If, however, he has an experimental and a control group, say, and if he has randomly assigned subjects to the two groups, he may assume that the effect of such variables is about the same on both groups—he may expect the two groups to differ on such variables only within the limits of random sampling. Hence, the extraneous variables should not differentially affect his dependent variable. And whatever the differences (small, we expect) between the groups on such variables, they are taken into account by our method of statistical analysis. For instance, the *t* test is designed so that it will tell us whether the groups differ on other than a chance basis.

The potential extraneous variables that might appear in the experimental situation are considerable. Various events might occur in an

[3] The inclusion of randomization as a technique of control is particularly arbitrary. This action may be defended on the grounds that the experimenter takes certain steps to insure its operation.

[4] An additional reason that randomization is important (such as random assignment of subjects to groups) is to insure the validity of the statistical test. But this point is covered later under the topic of assumptions of statistical tests (p. 285).

unsystematic way, such as the ringing of campus bells, the clanging of radiator pipes, peculiar momentary behavior of the experimenter such as a tic, sneezing, or scratching, an outsider intruding, odors from the chemistry laboratory, and the dripping of water from an overhead pipe. Now it might be possible to anticipate many of these variables and control them with one of our techniques, but even if it is possible, it might not be feasible. Signs may be placed to head off intrusions, but signs are not always read. A sound-deadened room is the answer to many of the problems, but such facilities are rare in psychological laboratories. It really must be assumed that *all* such variables will not be controlled by means of the previous techniques. Accordingly we can do the next best thing—we may assume that their effects will randomize out so that they will not differentially affect our groups. To facilitate the credibility of this assumption we might make sure that the order in which we run our subjects is approximately that of alternation. Thus, if the first subject we run is in the experimental group, the next might be in the control group, the third in the experimental group, etc. In this way we could expect, for example, that if a building construction operation is going on that is particularly bothersome, it will affect several subjects in each group and both groups approximately equally.

An Example of Exercising Extraneous Variable Control

To illustrate some of our major points, and to try to unify our thinking about control procedures, consider an experiment which has as its purpose the determination of whether the amount of stress present on members of a group influences the amount of hostility they will verbally express toward their parents while talking in that group situation. To answer this question we would first plan on collecting a number of individuals. Since we need to vary the amount of stress present on the members, we form two groups. A fairly heightened amount of stress is exerted on the experimental group (by some means that need not detain us here), while the control group experiences only the normal stress present in such a social situation. Our independent variable is thus amount of stress (which is varied in two ways), while the dependent variable is amount of hostility verbally expressed toward parents. Referring to Figure 6.4 we note that, as far as control is concerned, our first step is to determine the extraneous variables that are present. Through the procedures previously specified we might arrive at the following list: sex and age of subjects, whether their parents are living or dead, place of the experiment, time of day, characteristics of experimenter, lighting conditions, various noises, number in the groups, family background and ethnic origin of subjects, their educational level, recent experiences with parents, general aggressive tendencies, frustrations, previous feelings towards parents, weight, and height.

From Figure 6.4 we note that the next step is to determine those extraneous variables that might reasonably influence the dependent variable. Merely for illustrative purposes we have included two that probably will not influence the dependent variable and thus will be ignored: weight and height of subjects. All the rest must be dealt with. Those that might feasibly be controlled by elimination, holding conditions constant, or balancing, we might decide, are sex, age, place, time, lighting, group number, education, whether parents are living or dead, and experimenter characteristics. Of these, we could control the following by holding con-

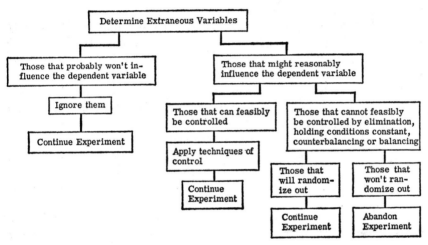

Figure 6.4. An over-all diagram of steps to be followed in planning an experiment.

ditions constant: place of experiment—by holding sessions for both groups in the same room; time of day—by holding sessions at the same time for both groups (on different days); lighting conditions—by having the same lights on for both groups with no external light present; number in the groups—by having the same number in each group; and experimenter characteristics—by having the same experimenter, with the same experimenter attitude, appear before both groups.

The variables of sex, age, educational level, and parents living or dead could be controlled by balancing. We could assign an equal number of each sex to each group, make sure that the average age of each group is about the same, distribute educational level equally between both groups, and assign an equal number of subjects whose parents are living to each group. Now, it is obvious that simultaneous balancing of all these variables would be rather difficult (if not impossible) with a small number of subjects. In fact, two variables would probably be as many as we could feasibly handle by this technique. We might select sex and parents, living or dead, as the most important and balance them out. For, if we are using college students (as we probably would), educational level and age would be about the same. Hence, we shall lump these two variables with the following as variables that we do not consider it feasible to control by the above techniques: various noises, family background, ethnic origin, recent experiences with parents, general aggressive tendencies, extent of frustration, and previous feelings toward parents. Some of these might be considered important and it would certainly be desirable to control them. Most of them are difficult to measure, however, and thus are hard to balance out in a specific manner.

Now, is it reasonable to assume that such variables will randomly affect both groups to approximately the same extent? If subjects are randomly assigned to groups (except insofar as we have restricted the random process through balancing), the assumption should be valid. And as previously noted we can always check on the validity of this assumption by comparing the groups on any given variable. Since this assumption seems reasonable to make in the present example we shall conclude that the conduct of the experiment is feasible as far as control problems are concerned—we have not been able to specify any confounding effects that would suggest that the experiment should be abandoned.

Some Control Problems

In the following experiments you should attempt to determine what the control problems are and to specify how you would apply the appropriate techniques to solve the problems. After considering the various experiments you should then reach a conclusion as to whether or not they should have been conducted.

To introduce this section we would like to discuss an experiment in which the control, if such existed, was outlandish. One day, a general called the author to say that he was repeating an experiment that the author had conducted on rifle markmanship, and asked if the author could visit him for the purpose of discussing the experiment. The trip was made and the general immediately drove out to the rifle range where the experiment was in progress. We visited the experimental group to observe their progress. During the visit it was more enjoyable watching the general than the subjects, who were newly "enrolled" army trainees. For while the trainees were practicing firing, the general would walk along the line, kicking some into the proper position, lying down beside others to help them fire, etc. After awhile the general suggested that we leave. That was fine, except for one thing, namely a desire to observe the control group. (By this time the author was beginning to wonder if there was a control group—but this concern was baseless). On expressing this desire, the general suggested a walk over the next hill, for that was where they were. On his way the author stopped to talk privately with the sergeant, particularly commenting on how lively and enthusiastic the experimental subjects were. The sergeant explained that that was what the general wanted—that the general expected the experimental group to fire better than the control group and they "darn" well knew that that was what had better happen. When the other side of the hill was reached the author was amazed at the contrast. For the control subjects were the most morose, depressed group of subjects he had ever seen. In talking to the sergeant in charge of this group he was informed that the general had never been to visit them. What is more, this group knew that they had better perform poorer than the experimental group—for nobody wanted the general to be disappointed (their motivations are too numerous to cite here). Needless to say, when the general reported the results of the experiment they were highly significant in favor of the experimental group.

1. The problem of whether children should be taught to read by the Word method or by the Phonics method has been a point of controversy for many years. Briefly, the Word method is where the child is taught to perceive the word as a whole unit, whereas the Phonics method requires that he break the word into parts. To attempt to decide this issue an experimenter plans to teach reading to two groups, one by each method. The local school system teaches only by the Word method. "This is fine for one group," he says. "Now I must find a school system that uses the Phonics method." Accordingly he leaves Wordville, and eventually finds another town that uses the Phonics method, Phonicsville. He then decides that he will test a sample of third-grade children in each town to see how well they can read. After administering a long battery of reading tests he finds that the children in Phonicsville are significantly superior to the children of Wordville. He then concludes that the Phonics method is superior to the Word method.

2. A military psychologist is interested in whether training to fire a machine gun from a tank facilitates accuracy in firing the main tank gun. He obtains a company of soldiers with no previous firing experience, and randomly divides them into two groups. One group receives .30 caliber machine gun training, the other does not. He then tests both groups on their ability to fire the larger tank gun. To do this he has two tanks set up so that they can fire on targets in a field. The machine-gun-trained group is assigned one tank and a corresponding set of targets,

while the control group fires on another set of targets from the second tank. His tests show that the group previously trained on the machine gun is significantly more accurate than the control group. His conclusion is that .30 caliber machine gun training facilitates accuracy on the main tank gun.

3. A psychologist seeks to test the hypothesis that early toilet training leads to a type of personality where children are excessively compulsive about cleanliness; conversely, late toilet training leads to sloppiness. The psychologist notes that previous studies have shown that middle-class children receive their toilet training earlier than do lower-class children. Accordingly he forms two groups, one of middle-class children and another of lower-class children. He then provides both groups with a finger painting task and records a number of data about their procedures, e.g., the extent to which they smear their hands and arms with paints, whether or not they clean up after the session, and how many times they wash the paints from their hands. Comparisons of the two groups on these criteria indicate that the middle-class children are significantly more concerned about cleanliness than are those of the lower-class. It is thus concluded that early toilet training leads to compulsive cleanliness whereas later toilet training results in less concern about personal cleanliness.

4. A physiological psychologist seeks to determine the function of the thalamus (an internal part of the brain). He obtains a sample of cats and randomly assigns them to two groups. An operation removes the thalamus from all the cats in one group. The second group is not operated on. On a certain behavior test it is found that the operated group is significantly deficient, as compared to the control group. The psychologist concludes that the thalamus is responsible for the type of behavior that is "missing" in the group that was operated on.

5. The following hypothesis is subjected to test: emotionally loaded words (e.g., sex, prostitute) must be exposed for a longer time to be perceived than words that are neutral in tone. To test this hypothesis various words are exposed to subjects for extremely short intervals. In fact, the initial exposure time is so short that no subject can report any of the words. The length of exposure is then gradually increased until each word is correctly reported. The length of exposure necessary for each word to be reported is recorded. It is found that the length of time necessary for subjects to report the emotionally loaded words is longer than for the neutral words. It is concluded that the hypothesis is confirmed.

7

The Independent and Dependent Variables

From one point of view, the primary purpose of an experiment is to test a hypothesis. And a hypothesis, we said, is a statement to the effect that two (or more) variables are related. We have referred to the two variables as the independent and the dependent variables. In this chapter we will discuss these variables in greater detail, and also the types of relationships that may obtain between them.

Types of Relationships Studied in Psychology

In general approach, we may develop an analogy between the way an engineer looks at a machine and the way that a psychologist looks at an organism. For the machine, the engineer first has to put some type of energy into it (he might call this "input"). The input then activates the machine in such a way that the energy is "carried" through it—this is the "throughput." And, finally, the machine accomplishes the task for which it is built and certain actions result, the "output." There are certain relationships among the input, throughput, and output so that for certain types or amounts of input, certain types and amounts of throughput occur, and certain types and amounts of output result. Furthermore, the characteristics of the machine limit the nature of the throughput and thus of the output. For example, only certain types or amounts of input are capable of being transmitted by the machine. And the characteristics of the machine determine what kinds of output may occur.

The psychologist's approach to behavior is analogous. For he may consider that the organism corresponds to the machine; the stimuli that excite the organism's receptors are the input, and the organism's responses are the output. The analogy may be pursued in the following manner: the type of stimuli that enter the organism (the input) de-

The Organism

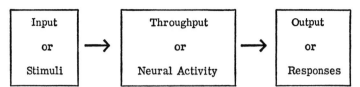

Figure 7.1. An analogy between the approach of an engineer to the study of a machine and the psychologist's approach to behavior.

termine what will happen within the organism (the throughput); and what happens within the organism influences the nature of the responses (the output). Furthermore, the specific characteristics of the organism, in particular the neural connections, also determine the nature of the organism's responses (see Fig. 7.1). For example, if a light is flashed in an organism's eye (input), various neural pathways are excited which go to and from the visual areas of the brain (throughput), and thence to specific effectors which result in a given response (output). However, in working with an organism whose visual areas in the brain have been destroyed, different characteristics will be encountered, and a different (or perhaps no) response would occur.

There are, then, three general classes of variables with which the psychologist deals: stimulus variables (the input), organismic variables (the throughput), and response variables (the output).[1] The psychologist attempts to determine relationships between these three. The possible relationships that may be studied are shown in Fig. 7.2.

These possible relationships may be indicated symbolically as follows:

1. $R_1 = f(R_2)$—response number one (any given response) is a func-

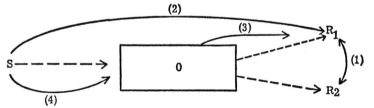

Figure 7.2. Showing the possible relationships among three classes of variables studied in psychology. S denotes the stimulus variables; O, organismic variables; R, response variables. The relationship indicated by numerals 1, 2, 3, and 4 are discussed in the text. (Modified from K. W. Spence, 1948.)

[1] Numerous classifications of variables with which psychologists deal are available elsewhere, e.g., Spence (1948) considers stimulus, organic, response, and hypothetical state variables; Underwood (1957) discusses environmental, task, instructional, and subject variables; Edwards (1950) uses the present classification as do Woodworth and Schlosberg (1955) though the latter add *antecedent variables* to the list.

tion of response number two (any other response).[2] When determining that two classes of responses are related, you are determining the first type of relationship. It is difficult to experimentally seek to establish a relationship between two responses. The application of correlational techniques are more appropriate to this problem. For example, an experimenter may want to know whether two dependent variables are related. In running rats in a maze, for instance, he wants to find out whether the number of errors that rats make is related to the total time it takes the rats to run the maze. These are two response measures and the correlation between them may be computed.

2. $R = f(S)$—a certain response class is a function of a certain stimulus (class). In this case, one may vary values of the stimulus to see if values of the response change. The stimulus may thus be seen to be the independent variable, while the response is the dependent variable. This second type of relationship is that with which we are most often concerned in experimentation. Probably the clearest examples of the areas of psychology in which this type of relationship is sought are those of perception and learning. In studies of perception we vary stimulus conditions and determine whether the perceptual responses of the organism also vary. For instance, we might vary the lighting conditions on a given object (varying the stimulus variable) and see if a person's verbal report of its size changes (a response measure).

3. $R = f(O)$—a response class is a function of (a class of) organismic variables. The primary purpose of research aimed at this type of relationship is to determine whether certain characteristics of the organism lead to certain types of responses. We might wonder, for instance, if people who are short and stout behave differently than people who are tall and thin. More specifically, do these two types of people differ as far as happiness, general emotionality, or degree of verbosity is concerned?

4. $O = f(S)$—a class of organismic variables is a function of a class of stimulus variables. In this case, we are primarily asking what environmental events influence the characteristics of organisms. For example, we might be interested in whether the degree to which a child has been isolated influences his intelligence.

This is a brief survey of the basic types of relationships sought by psychologists.[3] We should add that more complex relationships may also

[2] Let it be assumed that when we refer to a certain response we mean a certain response *class*, a number of quite similar instances of responses. For example, a response instance would be one hit at a baseball, whereas a response class would be made up of all of the times that we hit a baseball (a large number of response instances). Similarly, we may refer to stimulus instances and stimulus classes.

[3] You should note that these relationships conform to our discussion in Chapter 3 of the nature of hypotheses. There we showed how a hypothesis stated as a mathematical function [e.g., $R = f(S)$] is a special case of the more general "if *a* then *b*" relationship.

be sought, as for instance those that would occur if you investigate the relationship between three responses $[R_1 = f(R_2, R_3)]$, among two stimuli and a given response $[R = f(S_1, S_2)]$, or among a stimulus, response, and organismic variable $[R = f(O, S)]$. We would also note that the statement of the types of relationships sought depends on the way that you classify variables. Hence, other systems of classification would lead to different statements of the possible relationships.

The Independent Variable

In the first type of relationship we would vary one response to see if another is thereby affected. The response that we may vary would be our independent variable, the other response our dependent variable. However, as we said in presenting the first type of relationship, response-response relationships are not often sought with the use of standard experimental designs. In the second type of relationship we vary a stimulus and determine its effect on a response. Hence, the stimulus is the independent variable, the response the dependent variable. In the third type of relationship we vary an organismic variable as the independent variable and determine its relationship to a response, the dependent variable. And in the fourth type of relationship a stimulus is the independent variable and an organismic variable is the dependent variable. Thus, we have three independent variables to consider: responses, stimuli, and organismic variables. And in general, the symbol to the right of each equation is the independent variable, that to the left the dependent variable.

Response Variables

Because of the infrequent use of this type of independent variable in experimentation we shall not discuss it further, except to provide an example. Let us say that we are interested in whether people who read on the subway are steadier than people who do not read on the subway. It may have occurred to some of you who are both people watchers and subway riders that the reason that some people don't read on the subway is that they can't hold newspapers steady enough. Our procedure would be straightforward, assuming we could get the cooperation of our subjects: administer a suitable steadiness test to a group of subway readers and to a group of subway nonreaders. The difference, or lack of a difference, between the two groups would answer our question.

Stimulus Variables

Most independent variables with which we are concerned are stimulus variables, where the word stimulus is used in a broad sense to refer to any aspect of the physical or social environment. The following are examples of stimulus variables as they might affect a particular kind of behavior: the effect of different sizes of type on reading speed; the effect of different styles of type on reading speed; the effect of intensity of light on the rate of conditioning; the effect of number of people present at dinner on the amount of food eaten; the effect of social atmosphere on problem-solving ability. You will note that the variables differ considerably in their complexity.

Organismic Variables

By organismic variables we mean any relatively stable characteristic of the organism, including such physical or physiological characteristics as sex, eye color, height, weight, and body build, as well as such psychological characteristics as intelligence, educational level, anxiety, neuroticism, and prejudice.

At this point, let us remark that this system of classification has its disadvantages, since it is difficult to force some variables into categories. For example, we might question the placement of intelligence in the category of organismic variables. For while it can well be considered to be a characteristic of the organism, let us observe the way in which intelligence is frequently measured: a person takes a pencil and makes a number of marks on a piece of paper—he makes a number of responses. Hence, it is also possible to classify intelligence as a response variable.[4] We must, therefore, be quite arbitrary in some of our decisions, the justification being that we are trying to consider independent variables in an orderly fashion that will allow some systematic insight into the various kinds of variables used in psychological research.

A Further Consideration of Independent Variable Control

In Chapter 6 we said that independent variable control occurs when the researcher varies the independent variable in a known and specified manner. There are essentially two ways in which an investigator may exercise independent variable control: (1) purposive manipulation of

[4] To further complicate this decision, it would also be possible to consider intelligence as a logical construct, or in Spence's system of classification, as a hypothetical state variable (see note, p. 122).

the variable; and (2) selection of the desired values of the variable from a number of values that already exist. When purposive manipulation is used, we say that an experiment is being conducted; but when selection is used, we say that the method of systematic observation is being used. If an experimenter is interested in whether the intensity of a stimulus affects the rate of conditioning, he might vary intensity in two ways, high and low. If the stimulus is a light, he might choose the values, say, two and twenty candle power. He would then, at random, (1) assign his sample of subjects to two groups and (2) determine which group would receive the low intensity stimulus, which the high. In this case, he is *purposely manipulating* the independent variable (this is an experiment), for the decision as to what values of the independent variable to study and, more important, which group receives which value *is entirely up to him*. And, what is perhaps equally important, the experimenter himself "creates" the values of the independent variable.

Now, let us illustrate *independent variable control by selection of values* as they already exist (this is the method of systematic operation). Suppose that an investigator is interested in the effect of intelligence on problem-solving. Further, assume that he is not interested in studying the effects of minor differences of intelligence, but wants to study widely differing values of this variable. For example, he might wish to study three values: an IQ of 135, a second of 100, and a third of 65. Up to this point, the procedures in the two types of independent variable control are the same—the investigator determines what values of the variables he wants to study.

Now, in this case, the investigator must find certain groups that yield the values of intelligence that he wants. To do this he might administer a number of intelligence tests at three different places. First, he might select a number of bright college students, providing a group that has an average IQ of 135. Second, he might visit a rather nonselective group (high school students or army personnel) to obtain an average value of 100. And third, he might find a mental institution that would yield a group with an average IQ of 65. With these three groups constructed, he would then administer his test of problem-solving ability and reach his appropriate conclusion. Observe that he has selected the values of his independent variable from a large population. *The IQs of the people tested determined who would be the subjects. He has not, as in the preceding example, determined which subjects would receive which value of the independent variable.* Rather, in selection it is the other way around—the value of the independent variable determines which subjects will be used. It is thus apparent that in independent variable control, by selection of values as they already exist in subjects, *the subjects are not randomly assigned to groups.*

It is not really practical, however, to settle on precise IQ values, as above. What the experimenter is more likely to do is to say that "I want a very high intelligence group, a medium group, and a very low intelligence group." He might then make his three visits and settle for whatever IQs he gets—in this case, the averages might be 127, 109, and 72, which would probably still accomplish his purpose.

In short, *purposive manipulation* occurs when the investigator determines the values of the independent variable, "creates" those values himself, and determines which group of subjects will receive which value. *Selection* occurs when the investigator selects subjects who already possess the desired values of the independent variable.

The distinction between both kinds of independent variable control is important. To understand this, recall the intelligence-problem-solving example. What would be the investigator's appropriate conclusion? (Before hypothesizing possibilities, be sure to recall the chapter on control.) Consider the *confounded* nature of this investigation. We have three groups of subjects whom we know differ in intelligence. But in what other respects might they differ? The possibilities are so numerous that we shall only list three: socio-economic status, the degree of stimulation of their environments, and motivation to solve problems. Hence, whatever the results on the problem-solving tests, the confounding of our independent variable with extraneous variables would be atrocious. We would not know to which variable, or combination of variables, to attribute possible differences in dependent variable scores. This is not so with an experiment like the light-conditioning example above. For in that case, whatever the extraneous variables might be, they would be randomized out—distributed equally—over all groups.

When a stimulus variable is the independent variable, *purposive manipulation* is generally used. If the independent variable is either a response or organismic variable, however, *selection* is the more likely independent-variable-control procedure. For example, with intelligence (or number of years of schooling, or chronic anxiety, etc.) as the independent variable, we have no practical alternative but to select subjects with the desired values. It might be possible to manipulate purposively some of these variables, but the idea is impractical. It is admittedly difficult, say, to raise a person in such a way (manipulating his environment, administering various drugs, etc.) that he will have an IQ of the desired value; we doubt that you would try it.

A number of studies have been conducted to determine whether or not there is a relationship between smoking and lung cancer. Generally, two groups are studied, one composed of people who do not smoke, the second of those who do. The independent variable is thus the degree of smoking. Measures on the dependent variable are then taken—frequency of occurrence of lung cancer. The results have been generally decisive in that smokers more frequently acquire lung cancer than do nonsmokers. Nobody can argue with this statement. However, the additional statement is frequently made: *therefore,* we may conclude that smoking causes lung cancer. On the basis of the type of evidence presented above, such a statement is unfounded, for the type of control of the independent variable that has been used is that of selection of values. Numerous additional variables may be confounded with the independent variable.

The only way to determine here the cause-effect relationship is to exercise purposive manipulation control. That is, to select at random a group of subjects who have never been exposed to the smoking habit (e.g., children or isolated cultural groups), randomly divide them into two groups, and randomly determine which group will be smokers, which the abstainers. Of course, the experimenter must make sure that they adhere to these instructions over a long period. As members of the two groups acquire lung cancer, the accumulation of this evidence would decide the question. Unfortunately this experiment will probably never be conducted. In short, confounding is quite likely to occur when *selection* of independent variable values is used (the method of systematic observation) but is considerably less likely when purposive manipulation is resorted to (experimentation).

In studies involving more than one independent variable, the values of one variable might be purposively manipulated and the values of the other selected. Such an investigation may be referred to as a *quasi-experiment.*

The Dependent Variable

Measures of the Dependent Variable

Generally, we have viewed the response as the dependent variable in psychological experimentation. The definition of "response," as with our previous definition of "stimulus," is extremely broad. By "response" we mean to include such diverse cases as number of drops of saliva a dog secretes, number of errors a rat makes in a maze, time it takes a person to solve a problem, types of brainwaves (electroencephalograms), number of words spoken in a given period of time, accuracy of throwing a baseball, and judgments of people about certain traits. It is best to measure the response as precisely as possible. In many experiments great precision can be accomplished. We might briefly list some standard ways of measuring responses.

1. Accuracy of the response. Several ways of measuring accuracy are

possible. For example, we might have a metrical system, such as when we fire a rifle at a target. Thus, a hit in the bullseye might be scored a five, in the next outer circle a three, and in the next circle a one. Another type of response measure of accuracy is to count the number of errors the subject makes. For example, the number of erroneous movements a person makes in putting a puzzle together, or the number of blind alleys a rat enters in running a maze.

2. Latency of the response. The time that it takes the organism to begin the response is the latency. Reaction time studies are illustrative of this measure. The experimenter may provide a signal to which the subject must respond. He then measures the time interval between the onset of the signal and the onset of the response. Or, in the case of a rat running a maze, the latency might be the time interval between the raising of the start box door to the time the rat's hind feet are outside the door.

3. Speed of the response. This is a measure of how long it takes the organism to complete its response, once it has been started. If the response is a simple one like pressing one of two telegraph keys, the time measure would be quite short. But if it is a complex response, such as solving a difficult problem or assembling a complicated device, the time measure would be long. A measure of the speed of the response in the case of a rat running a maze would be the length of time between his leaving the start box, until he reaches the goal box. To distinguish between latency and speed measures—latency is the time between the onset of the stimulus to the onset of the response, while speed is the time between the onset and termination of the response.

4. Frequency of the response. One might also measure the number of times a response is made within a given period, actually a measure of the rate of responding. If a response is made ten times in one minute, the frequency of response is ten responses per minute. The rate gives an indication of the probability of the response—the higher the rate, the greater the probability that it will occur in the situation at some future time. The Skinner Box is the type of experimental situation in which frequency of response is most often used, since there the organism is usually allowed to make as many responses, and as frequently, as he "wishes."

Additional measures of the response might be level of ability that a subject can manifest (e.g., how many problems a subject solves with an *un*limited amount of time), or the intensity of a response (e.g., the volume of salivation in a conditioning study). Frequently it is impossible to obtain an adequate measure of the dependent variable with any of the above techniques. In this event, it might be possible to devise a rating scale. For example, a rating scale for anxiety might have five gradations: 5 meaning "extremely anxious," 4 "moderately anxious," and so on. Competent judges would then mark the appropriate position on the scale for each subject. Or the subjects could even rate themselves.

Objective tests frequently serve as dependent variable measures (these usually would measure what we have called organismic variables). For example, you might wish to know whether psychotherapy decreases a person's neurotic tendencies, in which case, you might administer a standard test for this purpose. If a suitable standard test is not available, it might be that you could construct your own.

These are some of the more commonly used measures of dependent variables. It is impossible to cover every situation with precision. By combining some of the above ideas with your own ingenuity, you should be able to arrive at an appropriate measure of a dependent variable for the independent variable that you wish to study.

Selecting a Dependent Variable

The experimenter seeks to determine if his independent variable affects a dependent variable. But how does the experimenter determine what dependent variable to measure and record? Behavior is exceedingly complex. At any given time an organism makes a fantastically large number of responses. Take Pavlov's simple conditioning experiment with dogs. His dependent variable (as it is most frequently cited) was amount of salivation. However, that is not the dog's only response when a conditioned and an unconditioned stimulus are simultaneously presented. For, in addition to salivating, he also breathes at a certain rate, wags his tail, moves his legs, pricks up his ears, and so on. Now, out of this mass of behavior, Pavlov had to select a particular response to measure and record. He might have studied some response other than salivation, a response that might or might not have also been related to his independent variable. Why did he choose salivation?[5]

Presumably every stimulus-independent variable leads to certain responses. The problem of selecting a dependent variable, then, would seem simply to find all the responses that are influenced by a given stimulus-independent variable. But the problem is not quite that simple, and even this answer is not a simple one to follow in practice. Look at the matter in terms of our previous distinction between exploratory and confirmatory experiments. In the exploratory experiment, the experimenter asks himself: "I wonder what would happen if I did this?" He selects some responses to measure to see if they are affected by the independent variable. For it is impossible in any practical sense to study all of them—you simply pick and hope. An interesting example of this procedure is offered in the following quotation:

[5] See Skinner (1953, pp. 52-54) for a brief, interesting discussion of this point.

". . . The discovery that serotonin is present in the brain was perhaps the most curious turn of the wheel of fate. . . . Several years ago Albert Hofman, a Swiss chemist, had an alarming experience. He had synthesized a new substance, and one afternoon he snuffed some of it up his nose. Thereupon he was assailed with increasingly bizarre feelings and finally with hallucinations. It was six hours before these sensations disappeared. As a reward for thus putting his nose into things, Hofman is credited with the discovery of lysergic acid diethykanide (LSD), which has proved a boon to psychiatrists because with it they can induce schizophrenic-like states at will . . ." (Page, 1957, p. 55).

In this "pick and hope" procedure you can reach two possible conclusions from your data: (1) the independent variable did not affect the particular variable; or—(2) it did. In the confirmatory experiment, on the other hand, you have a precise hypothesis that indicates the dependent variable in which you are interested—it specifies that a certain independent variable will influence a certain dependent variable. Your procedure is straightforward, at least in principle. You merely select a measure of the dependent variable in question and see if your hypothesis is probably true or false.

Since the dependent variable has already been specified by the hypothesis, you must be careful to obtain a proper measure of it. That is, you must be sure that the data you record are actually measures of the dependent variable *in which you are interested.* Suppose, for instance, that instead of measuring amount of salivation Pavlov measured the change in color of his dog's hair—the whole concept of conditioning would then have been delayed until the appearance of a shrewder investigator. This is a grotesque example, but more subtle errors of the same type are frequently made in psychological research. One experimenter wishing to study the effect of a certain independent variable on emotionality might select several judges to rate the apparent emotionality of his subjects after the independent variable has been introduced. Whatever his results, he should ask the question: did the judges actually rate subjects on the basis of emotionality, or did they unknowingly rate them on some other characteristic? It may have been that the subjects were actually rated on "general disposition to life," "intelligence," "personal attractiveness," or whatever. If this actually happened, we could not say that emotionality was the dependent variable. This brings us to the first requirement that a dependent variable in a confirmatory experiment must meet—it must be *valid.* The data recorded must reflect measures of the characteristic that the experimenter seeks to measure.

"Now," you might say recalling our discussion on operational definitions, "if the experimenter defined emotionality as what the judges reported, then that is by definition emotionality—you can't quarrel with that." And so we can't, at least on the grounds that you offered. For we recognize that anyone can define anything anyway that he wants. You can, if you wish, define the typical four-legged object with a flat surface on top from which we eat "a chair" if you like. Nobody would say that you can't. However, at the same time, we must ask you to look at a social criterion: is that the name usually used by other people? Obviously the object is usually referred to as a "table." And if you insist on referring to what the rest of us call a "table" as a "chair" nobody should call you wrong, for that is your privilege. However, you will be at a distinct disadvantage when you try to communicate with other people. When you invite your dinner guests to be seated on a table and to eat their food off of a chair some very quizzical responses will be evoked to say the least.

So the moral is this: while you may define your dependent variable as you wish, it is wise to define your dependent variable as it is customarily used—at least, if it *is* customarily used. And if you are lucky enough to investigate a problem that has a certain widely accepted definition for the dependent variable involved, you should either use that dependent variable or one that correlates highly with it.

Consider dependent variable validity by some additional examples. Suppose you are interested in determining the influence of a given independent variable on problem-solving. You might define your dependent variable as the number of problems of a certain nature solved within a given period of time. At first glance, this *seems* a fine example of a valid dependent variable. And, if the test has a large number of problems and if these problems are arranged in ascending order of difficulty then it probably is valid. But if the test is lengthy and the problems are all easy, you probably would not be measuring problem-solving ability but, rather, reading speed. That is, regardless of the fact that "problems" are contained in the test, those who read fast would get a high score and those who read slowly would get a low score. Clearly this would not be a valid measurement of problem-solving ability (unless, of course, problem-solving ability and reading speed are significantly correlated). Or, to make the matter even simpler, if you construct a very short test composed of extremely easy problems, all the subjects would get a perfect score, unless you are working with feebleminded individuals or the like. This test is not a valid measure of the dependent variable.

In the example where we were interested in whether rats could learn to run to a white or black goal box, the training procedure consisted of feeding them in a certain goal box (say a white one). The test consists of running the rats for a number of trials in a maze which contains one black and one white goal box. Let us say, for the purposes of making our point, that the white box is always on the right and the black box is always on the left. Assume that the preponderance of runs we record are to the white box. We conclude that rats run to the box of the color in which they were previously fed. Now, are we really measuring the extent to which they run to the white box? Rats have position habits— they frequently are either "left turners" or "right turners" (or they may alternate in certain patterns). If we have selected a group of rats that are all right turners, our measure may be simply of position habits, rather than of the dependent variable we are interested in.

To determine the validity of the dependent variable, the experimenter might correlate his dependent variable scores with scores obtained by the same subject on some other measure that is known to be valid. If the correlation is high his measure is valid; if low or not significant, it is not valid. For example, suppose we have a valid measure of anxiety available, and that we wish to use a different measure of anxiety as our dependent variable. To determine the validity of the measure that we wish to use, we would take both measures on a number of subjects. By computing the correlation between the two sets of scores, we could reach a conclusion as to the validity of our measure. Unfortunately, such a procedure is seldom practical in an experimental situation. For one, we generally do not have a measure of the desired dependent variable that has known validity. Furthermore, this approach to validity is limited. The whole concept of validity is currently undergoing considerable re-examination (e.g., Bechtold, 1959; Loevinger, 1957; Cronbach & Meehl, 1955). We must still await final answers about kinds of validity and the operational techniques for measuring them (especially as they apply to experimentation). We have attempted to point out some of the types of problems present, and some of the kinds of pitfalls that await unwary experimenters. At present, the best that we can say is to be aware of these problems and potential errors; then, on the basis of considerable reflection of the nature discussed above and on the basis of results of previous research, select the measure of your dependent variable that appears to be the most valid.

The second requirement that a dependent variable should satisfy is that of *reliability* (in part) the extent to which subjects receive the same scores when repeated measurements are taken. For example, an intelligence test may be considered sufficiently reliable if subjects make approximately the same scores every time they take the test. Suppose a subject received an IQ of 105 the first time he takes a test, 109 the second, and 102 the third. These scores are approximately the same. If all subjects behaved similarly, it may be said that the test is reliable for the population sampled. However, suppose that the subject scored 109, 138, and 82. Such a test could not be considered reliable, for the repeated measurements vary too much.

Considering an experimental situation, reliability of a dependent variable could be measured in the following manner, although alternative approaches might be used. First, the experimenter could obtain measures of his subjects on the dependent variable. After a period of time, he could obtain measures on the same subjects on the same measure again. He would then compute the correlation between the two sets of measures. If the correlation is high, his dependent variable measure is reliable; otherwise it is not.

We said above that reliability is in part the extent to which subjects receive the same scores when repeated measurements are taken. To elaborate on this, let us note that experimenters frequently study subject's characteristics that change with the passage of time. They may study a learning process or the growth of anxiety. In such a case we need merely note that the correlation of successive scores may be quite high providing that the subjects maintain the same relative scores: That is, providing that the rank order of scores is similar on each testing. For example, if three subjects made scores of 10, 9, and 6 on the first testing, but 15, 12 and 10 respectively on the second testing, the correlation (reliability) would be high since they maintained the same relative ranks. If, however, the first subject's score changed from 10 to 11, the second from 9 to 12, and the third from 6 to 12, the reliability would be lower. In short, then, whether or not the measures of the dependent variable change with time, a correlation coefficient can be computed to determine the extent to which the dependent variable is reliable.

Unfortunately, most experimenters do not bother to determine the reliability of their dependent variables. Our knowledge about experimentation on problems would be increased if this were the case. At the same time we must point out that the determination of reliability is sometimes unrealistic. For there are situations in which the dependent variable is more reliable than a computed correlation coefficient would indicate. For one thing, the subjects used are frequently too homogeneous to allow the computed correlation value to approach the true value. For instance, if all subjects in a learning study had precisely the same ability to learn the task presented them, then (ideally) they would all receive exactly the same dependent variable scores on successive testings—the computed correlation would be zero when in fact the dependent variable might be quite reliable. Another reason for a lower computed correlation than the true one is that the scale used to measure the dependent variable has insufficient range. To illustrate again by taking an extreme case, suppose that only one value of the dependent variable was possible. In this event all subjects would receive that score, say five, and the computed correlation would again be zero. More realistically an experimenter might use a five point scale as a measure of his dependent variable, but the only difference between this and our absurd example is one of degree —the five point scale might still be too restrictive in that it does not allow one to sufficiently differentiate among the true scores of his subjects; two subjects for instance might have true scores of 3.6 and 3.9, but on a scale of five values they would both receive scores of 4.0.

Recognizing that it is desirable for the experimenter to determine the reliability of his dependent variable and that it is frequently not feasible to approximate the true value by correlating successive scores, we may ask what the experimenter does. The answer is that he plans the experiment and collects his data. If he finds that his groups differ significantly, he may look back and reach some tentative conclusions about the reliability of the dependent variable. For if his groups differ significantly, this means that they differ on other than a chance basis. And, if the means on his dependent variable have differed in other than a chance manner, it must possess sufficient reliability—for lack of reliability makes for only chance variation in scores. On the other hand, if his groups do not differ significantly, this means that the scores are probably due to chance variation. The conclusion that would most frequently be reached in such a situation is that variation of the independent variable does not affect the dependent variable. Other reasons are also possible. It may be that the dependent variable is unreliable. So this approach to determining reliability is a one-way affair: if there are significant differences among groups, the dependent variable is probably reliable; if there are no significant differences, then no conclusion about reliability is possible (at least on only this information). The repetition of the experiment a number of times with consistently significant results would increase our belief in the reliability of the dependent variable. For if the same results are continually obtained, certainly the dependent variable is reliable.

The concepts of validity and reliability have been extensively used by test constructors. They have been almost totally ignored by experimenters, and yet their great importance to experimentation should be apparent. If the experimenter has not selected a valid and reliable criterion (dependent variable), his experiment is worthless. If he is performing a learning experiment and his dependent variable actually measures drive, then obviously his conclusions with regard to learning are baseless. The close tie between validity of the dependent variable and experimental control should be apparent. If an uncontrolled extraneous variable is affecting the dependent variable, the measures obtained may be valid for the extraneous variable, but not for the independent variable. On the other hand, if the learning experiment has an unreliable dependent variable, then the scores of the subjects, regardless of the experimental conditions, would vary at random. With all of the subject's scores varying in a chaotic manner, it is impossible to determine the effectiveness of the independent variable.

Multiple Dependent Variables

A given independent variable may affect a number of measures of behavior. Such a number of measures is frequently recorded. For example, an experiment in which rats are run through a maze might use all of the following measures of behavior: time that it takes to leave the starting box (latency), time that it takes to enter the goal box (running time or speed), number of errors made on each trial, and number of trials required to learn to run the maze with no errors. Such an experiment could be looked upon as one with four dependent variables, in which case the experimenter would merely conduct four statistical analyses; he might run, say, four separate *t* tests to see if his groups differ on any of the four dependent variables.

An alternative procedure might be to obtain the correlations between the various dependent variables under consideration for use. These correlations might be based either on previous research or on the results of a pilot study. You might find that two potential dependent variables correlate quite highly, e.g., .90.[6] In this case, you would know that they are measuring largely the same thing, and there would be little point in recording both measures. Hence, you would select one or the other, probably the easiest to record. We must emphasize, however, that the correlations between your dependent variables should be quite high before you eliminate any of them. The author once conducted an experiment in which three dependent variables were used. It was found that the correlation between the first and the second was .70, between the first and the third .60, and between the second and the third .80. Yet, in spite of these high correlations, the statistical analysis indicated that there was no difference between experimental conditions on two of the dependent variables, but that there was a difference significant beyond the 1 per cent level for the conditions on the third criterion.

In short, then, it is desirable to measure every dependent variable that might reasonably be affected by your independent variable. This statement is offered as an ideal that can only be approached, for clearly in any actual experiment, it would not be feasible. If you have definite information that indicates a high correlation among some of your dependent variables, you might choose among them.

Growth Measures

More often than not in psychology experimenters deal with variables that change with time. This is almost universally true with learning studies. For example, we may be interested in how a skill grows with

[6] See p. 142 for a discussion of correlation and an interpretation of this number.

repeated practice under two different methods. Frequently a statistical test is run on terminal data, i.e., data obtained on only the last trial. However, the learning curves of the two groups could provide considerable information about how the two methods led to their terminal points —one method might have been a slower starter but gained more rapidly at the end. And, in addition to providing such valuable information you may want to run statistical tests which compare the two curves at specific points or even as whole units.[7]

Delayed Measures

Another important question concerns the possible retention of experimental effects. For example, we might find that one method leads to better learning than a second method, but we might also be interested in whether that advantage is maintained over a period of time (with various kinds of intervening activity). Suppose that your task is to train mechanics in the performance of a highly technical job. The training you give them is to be followed by training on something else, so that it will be quite a while before they will actually use the training they received from you. In this case, you would not only be interested in which of several methods is more efficient for learning, but also which method leads to the best retention. If you conducted an experiment, you might have the men return to you for another test just before they started their on-job duty. On the basis of this delayed test, you could decide which of the several methods would be best to use. Unfortunately, psychologists seldom take delayed measurements of their experimental effects, even when such a practice would be quite easy for them.

[7] This is known as trend analysis, see for example Lindquist (1953).

8

Experimental Design: The Case of Two-Matched-Groups

The type of design that we have considered up to this point is the two-groups design which requires that subjects be assigned to each group in a random fashion. This design is based on the assumption that chance assignment will result in two essentially equal groups (as determined, of course, by comparing their means.) The validity of this assumption increases with the number of subjects used.

The basic logic of all experimental designs is the same: start with groups that are essentially equal, administer the experimental treatment to one and not the other, and note the changes on the dependent variable. If the two groups had equal means on the dependent variable before the administration of the experimental treatment, and if a significant difference between the means of the groups on the dependent variable results after the administration of the experimental treatment, and if extraneous variables have been adequately controlled then that difference on the dependent variable may be attributed to the experimental treatment. The matched groups design is simply one way of helping to meet the assumption that the groups have essentially equal dependent variable scores prior to the administration of the experimental treatment.

An Example of a Two-Matched-Groups Design

Let us say that we are interested in testing the hypothesis that both reading and reciting material leads to better retention than reading alone. We might form two groups of subjects, one to learn the material by reading and reciting, the second group to spend all their time in reading. If we were using a randomized groups design we would assign subjects to the two groups at random. With the matched groups design, however, we assign them according to scores on an initial measure called the matching variable. Let us say that we use intelligence test scores as our matching variable. We might have ten subjects available; their scores are indicated in Table 8.1.

Table 8.1

Scores of a Sample of Subjects on a Matching Variable

Subject	Intelligence Test Score
1	120
2	120
3	110
4	110
5	100
6	100
7	100
8	100
9	90
10	90

Our strategy is to construct the groups so that they are equal in intelligence. To accomplish this we need to pair the subjects who have equal scores, assigning one member of each pair to each group. It is apparent that the following subjects can be paired: 1 and 2, 3 and 4, 5 and 6, 7 and 8, and 9 and 10. The method we shall use for dividing these pairmates into two groups is randomization. This assignment by randomization is necessary in order to prevent possible experimenter biases from interfering with the matching. For example the experimenter may, even though he is unaware of such actions, assign more highly motivated subjects to one group even though each pair has the same intelligence score. By a flip of a coin we might determine that subject 1 goes in the Reading and Reciting group, number 2 goes in the Reading group. The next coin flip might determine that subject 3 goes into the Reading group and number 4 in the Reading and Reciting group. And so on for the remaining pairs (see Table 8.2).

Table 8.2

The Construction of Two Matched Groups on the Basis of Intelligence Scores

READING GROUP		READING AND RECITING GROUP	
Subject	Intelligence Score	Subject	Intelligence Score
2	120	1	120
3	110	4	110
6	100	5	100
7	100	8	100
10	90	9	90
	———		———
	520		520

We may note that the sums (and therefore the means) of the intelligence scores of the two groups in Table 8.2 are equal. Let us now assume that the two groups are subjected to their respective experimental procedures and that we obtain the retention scores for them indicated in Table 8.3 (the higher the score, the better they retained the learning material). We may note that we have placed the pairs in rank order according to their initial level of ability on the matching variable, i.e., the most intelligent pair is placed first, while the least intelligent pair is placed last.

Table 8.3

Dependent Variable Scores for Pairs of Subjects Ranked
on the Basis of Matching Variable Scores

INITIAL LEVEL OF ABILITY	READING GROUP		READING AND RECITING GROUP	
	Subject	*Retention Score*	*Subject*	*Retention Score*
1	2	8	1	10
2	3	6	4	9
3	6	5	5	6
4	7	2	8	6
5	10	2	9	5

Statistical Analysis of a Two-Matched-Groups Design

Observation of the two groups of scores in Table 8.3 suggests that the group that both read and recited their material is superior to the reading-only group, but as before, we must ask, are they significantly superior? To answer this question we may apply the *t* test, although the application will be a bit different for a matched groups design. The equation is:

$$(8.1) \qquad t = \frac{\overline{X}_1 - \overline{X}_2}{\sqrt{\dfrac{\Sigma D^2 - (\Sigma D)^2/n}{n(n-1)}}}$$

The symbols are the same as those previously used, with the exception of D, which is the difference between the dependent variable scores for each pair of subjects. To find D we subtract the retention score for the first member of a pair from the second. For example, the first pair consists of subjects 2 and 1. Their scores were 8 and 10, respectively, and $D = 8 - 10 = -2$. Since we will later square the D scores [to obtain $(\Sigma D)^2$], it makes no difference which group's score is subtracted from which. We could just as easily have said: $D = 10 - 8 = 2$. The only caution to observe is that we need to be consistent, i.e., we must always subtract the Reading group's score from the Reading-Reciting group's score, or vice versa. Completion of the D calculations is shown in Table 8.4.

Table 8.4

Computation of the Value of D

Initial Level of Ability	Reading Group	Reading and Reciting Group	D
1	8	10	−2
2	6	9	−3
3	5	6	−1
4	2	6	−4
5	2	5	−3

Our formula instructs us to perform three operations with respect to D. First, to find the sum of the squares of D (ΣD^2) and second, to find the sum of D (ΣD) and third, to compute $(\Sigma D)^2$. ΣD^2 instructs us to square each D value and sum them, i.e., $(-2)^2 + (-3)^2 + (-1)^2 + (-4)^2 + (-3)^2 = 4 + 9 + 1 + 16 + 9 = 39 = \Sigma D^2$. ΣD tells us to sum D; i.e., $(-2) + (-3) + (-1) + (-4) + (-3) = -13 = \Sigma D$. And of course $(\Sigma D)^2$ is the square of ΣD, i.e., $(\Sigma D)^2 = \Sigma D \times \Sigma D$. n we may recall is the number of subjects in a group (not the total number of subjects in the experiment). In the matched groups design we may assume that the number of subjects in each group is the same, otherwise they would not be matched.[1] In our example $n = 5$. The numerator is the difference between the (dependent variable) means of the two groups, as with the previous use of the t test. The means of the two groups are 4.6 and 7.2. Substitution of these values in equation 8.1 results in the following:

$$ t = \frac{7.2 - 4.6}{\sqrt{\dfrac{39 - (-13)^2/5}{5(5-1)}}} = 5.10 $$

The equation for computing the degrees of freedom for the matched t test is $df. = n - 1$. (Note that this is a different equation from that for the two-randomized-groups design.) Hence, for our example, $df = 5 - 1 = 4$. Consulting our Table of t, as before, with a t of 5.10 and 4 degrees of freedom we find that our t is significant beyond the 1% level ($P < 0.01$). We may thus reject our null hypothesis (that there is no difference between the population means of the two groups) and conclude that the groups differ significantly. Since our mean for the Reading-Reciting group is higher than that for the Reading group we may conclude

---

[1] For an exception see Peters and Van Voorhis (1940, p. 463).

that the former is significantly superior and that our hypothesis is confirmed.

Correlation and the Two-Matched-Groups Design

To adequately understand this design we shall have to consider the topic of correlation. A correlation is a measure of the extent to which two variables are related. The measure of correlation that we shall be concerned with is symbolized by r. This symbol stands for the correlation coefficient (more completely, the Pearson Product Moment Coefficient of Correlation). Since the value of r tells us the extent to which two variables are (linearly) related, it is a very valuable statistic that is used in a variety of ways. There is an equation for computing r but we shall not discuss it here since we are only interested in certain principles.[2] The value that r may assume varies between $+1.0$ and -1.0. A correlation of $+1.0$ is a perfect positive correlation. To illustrate this let us say that a group of people have been administered two different intelligence tests. Since both tests presumably measure the same thing we may assume that the scores are highly correlated. They might be as indicated in Table 8.5.

Table 8.5

Scores on Two Intelligence Tests Received by Each Subject

Subject	Score on Intelligence Test A	Score on Intelligence Test B
1	120	130
2	115	125
3	110	120
4	105	115
5	100	110
6	95	105

We may note that the subject who received the highest score on test A also received the highest score on test B. And so on down the list, subject 6 receiving the lowest score on both tests. A computation of r for this very small sample would yield a value of $+1.0$. Hence, the scores on the two tests are perfectly correlated—whoever is highest on one test is also highest on the other test, whoever is lowest on one is lowest on the

[2] A straight forward equation for computing r directly from original data is:

$$r_{XY} = \frac{n\Sigma XY - (\Sigma X)(\Sigma Y)}{\sqrt{[n\Sigma X^2 - (\Sigma X)^2][n\Sigma Y^2 - (\Sigma Y)^2]}}$$

r_{XY} is the correlation between two variables X and Y and ΣXY is the sum of the cross products of the values of X and Y for each subject.

other, and so on *with no exception being present*.[3] Now let us say that there are one or two exceptions in the ranking of the test scores. Suppose that subject 1 had the highest score on test A but the third highest score on test B; that number 3 had the third highest score on test A but the highest score on test B; and that all other relative positions remained the same. In this case the correlation would not be perfect (1.0) but would still be rather high (it would actually be .77).

Moving to the other extreme let us see what a perfect negative correlation would be, i.e., one where $r = -1.0$. We might administer two tests, one of democratic characteristics and one that measures amount of prejudice (see Table 8.6). The person who scores highest on the first test

Table 8.6

Scores on Two Personality Measures for Each Subject

Subject	Score on Test of Democratic Characteristics	Score on Test of Prejudice
1	50	10
2	45	15
3	40	20
4	35	25
5	30	30
6	25	35

receives the lowest score on the second. This inverse relationship may be observed to hold for all subjects without exception, resulting in a computed r of -1.0. Again if we had one or two exceptions in the inverse relationship the r would be something like $-.70$ indicating a negative relationship between the two variables, but one short of being perfect.

To summarize, given measures on two variables for each subject a positive correlation exists if, as the value of one variable increases, the value of the other one also increases. If there is no exception the correlation is perfect; if there are relatively few exceptions it is positive but not perfect. Thus, as test scores on intelligence test A increase the scores on test B also increase. On the other hand, if the value of one variable decreases while that of the other variable increases, a negative correlation exists. No exception indicates that the negative relation is perfect; otherwise it is less than perfect. Hence as the extent to which a person exhibits democratic characteristics increases the amount of his prejudice decreases; and this, of course, is what we would expect.

---

[*] Actually another characteristic of the scores must also be present for this type of correlation to be perfect. That is that the interval between the scores of each set must be proportional. In our example 5 IQ points separate each subject on each test. However, this characteristic is not crucial to the present discussion and will not be considered further.

One final point concerning the value that r may assume. If $r = 0$ one may conclude that there is a total lack of (linear) relationship between the two measures. In other words as the value of one variable increases the value of the other variable varies in a random fashion. Examples of situations where we would expect r to be zero would be where we would correlate hair color with intelligence, or number of books that a person reads in a year with the length of his toenails. Additional examples of *positive* correlations would be the height of a person and his weight, his IQ and his ability to learn, and his grades in college and his high school grades. We would expect to find negative correlations between the amount of heating fuel a family uses and the outside temperature, or the weight of a person and his success as a jockey.

In science we seek to find relationships between variables. And a negative relationship (correlation) *is just as important* as a positive relationship. Do not think that a negative correlation is undesirable or that it indicates a lack of relationship. To illustrate, a correlation of $-.50$ indicates just as strong a relationship as a correlation of $+.50$, and a correlation of $-.90$ indicates a stronger relationship than does one of $+.80$.

Selecting the Matching Variable

Recall that in matching subjects we have attempted to equate our two groups with respect to their mean scores on the dependent variable. In other words, we have selected some initial measure of ability by which to match the subjects and have assigned them to two groups on the basis of these scores so that the two groups are essentially equal on this measure. If the matching variable is highly correlated with the dependent variable scores, our matching has been successful.[4] For in this event we largely equate the groups on their dependent variable scores by using the indirect measure of the matching variable. If the scores on the matching variable and the dependent variable do not correlate, however, then our matching is not successful. In short, the degree to which the matching variable scores and the dependent variable scores correlate is an indication of our success in matching.

How can we find a matching variable that correlates highly with our

---

[4] Let us emphasize that r is a measure of the extent to which two variables (in our case the matching and the dependent variables) are *linearly* related. We are, thus, simplifying our discussion by considering only linear relationships between our two variables. The justification for excluding curvilinear relationships from the above discussion can also be offered that our knowledge about the several possible correlations involved in equation 8.3 is considerably limited.

dependent variable? It might be possible to use the dependent variable itself. For example, if we are studying the process of learning to throw darts at a target, there might be two methods of throwing that we wish to evaluate. The design would call for one group of subjects to use method A, the other method B. To assign subjects to the groups by matching we might first have all subjects throw darts for five trials. We could use their scores on these five trials as the basis for pairing them off into two groups. They would then be trained by the two methods, respectively, and a later proficiency score computed. The *t* test for matched groups would then be run on that later proficiency score. Our matching would be judged successful to the extent that the first set of scores correlated with the later set of scores. Since both sets of scores are obtained on the same task we would expect the correlation to be rather high. Thus, it is clear that *the matching variable that is most likely to show a correlation with the dependent variable is that dependent variable itself.*

However, it should be apparent that this technique is not always feasible. Suppose that the dependent variable is a measure of rapidity in solving a problem. If practice on the problem is first given to obtain matching scores, then everyone would know the answer when it is administered later as a dependent variable. Or, take another example, where we create an artificial situation to see how people react in a panic. Using the same situation to take initial measures for the purpose of matching subjects would destroy the novelty of the situation after the independent variable is administered. In such situations we must find other measures that are highly correlated with dependent variable performance.

In the problem solving example we might give the subjects a different, but similar problem to solve, and match on that. Or, if our dependent variable is a list of problems to solve, we might select half that list to use as a matching variable and use the other half as a dependent variable. In the panic example it might be reasonable to assume that a physiological measure of stress would be related to performance during a panic. For example we might take a measure of how much the subjects sweat under normal conditions and assume that those who normally sweat a lot are highly anxious individuals (cf. Mowrer, 1953, pp. 591-640). Matching on such a test might be feasible.

Perhaps the most widely used matching variable in learning studies is a measure of intelligence. The assumption is that the higher the intelligence the better the learning capacity. Intelligence test scores are quite easy to obtain, or may perhaps already be on file in the case of college students.

Another general possibility is to match subjects on more than one variable. In a learning experiment, we might match subjects on initial learning scores *and* intelligence. Further consideration might suggest additional measures that could be combined with these two.[5]

Now we have said that if the matching variable does not correlate rather highly with the dependent variable, a matched-groups design should probably not be used. For this reason you should be rather certain that a high correlation exists between both variables. You might consult previous studies, for they may provide information on correlations between your dependent variable and various other variables. You could then make a selection from among those that correlate most highly. Of course, you should be as sure as possible that the same correlation holds for your subjects with the specific techniques that you use.

You might also conduct a pilot study where you would take a number of measures on some subjects and also administer your dependent variable to them. Selection of the measure with the highest correlation with the dependent variable would afford a fairly good criterion, if it is sufficiently high. If it is too low, you should pursue other possibilities, or consider abandoning the matched-groups design.

One procedural disadvantage of matching occurs in many cases. When using initial trials on a learning task as our matching variable, we need to bring the subjects in to the laboratory to obtain the data on which to match them. Then, after computations have been made and the matched groups formed, the subjects must be brought back for the administration of the independent variable. The requirement that subjects be present twice in the laboratory is sometimes troublesome. It is more convenient to use measures that are already available, such as intelligence test scores or college board scores. It is also easier to administer group tests, such as intelligence or personality tests, which can be accomplished in the classroom. On the basis of such tests appropriate subjects can be selected and assigned to groups before they enter the laboratory.

A More Realistic Example

The example of a matched-groups design and its statistical analysis that we previously used was constructed so that we could "breeze

---

[5] However it is frequently advisable to use a special technique for combining the various measures, discussion of which is probably not too fruitful here. For further information on one way to use more than one matching variable you are referred to Peters and VanVoorhis (1940).

through" it in order to observe the general principles involved. There are, however, a number of details that prove somewhat troublesome when using this design, so let us illustrate it with another, more realistic problem.

To consider a previous example let us say that the government's Civil Defense organization is interested in whether specific training leads people to behave more efficiently when a disaster occurs. To answer this question our design would call for the construction of two matched groups, one group to receive the panic training, the other group to receive no training. We might bring these people into a room and excuse ourselves under some pretense. There could then occur in the room a minor explosion followed by the release of considerable smoke and perhaps even some small flames. Our dependent variable would be how fast they could find a rather obscure exit from the room. To form our two matched groups we would obtain measures of sweat on all subjects to use as their initial anxiety scores (see Table 8.7).

Table 8.7

The Use of Anxiety Scores as a Matching Variable

Subject Number	Anxiety Score
1	20
2	14
3	13
4	13
5	12
6	9
7	9
8	9
9	9
10	9
11	8
12	6
13	3
14	2
15	1
16	0

We now have measures on sixteen subjects from which to construct two matched groups. The pairing of subjects is not as easy as with the previous example. Who, for instance, would we pair with subject 1 who has such a high score of 20? In fact, there are few subjects who have precisely the same scores, so it appears we are going to have to compromise. We could pair the subjects who have scores of 20 and 14 together, saying this is the best match that we can make. Or we could exclude the first subject from our sample, but that would also necessitate excluding another subject since we require two groups equal in number. After considerable soul searching we might finally arrive at the division of subjects in Table 8.8.

Table 8.8

Matching Subjects on the Basis of Anxiety Scores
and Randomly Assigning Pairmates to Two Groups[6]

Experimental Group	*Control Group*
14	13
12	13
9	9
9	9
9	8
2	3
0	1
$\overline{X} = 7.86$	$\overline{X} = 8.00$
$s = 5.08$	$s = 4.58$

By excluding subjects number 1 and 12 we have been able to achieve a fairly good matching between the two groups, as seen by comparing their means.[7] Now assume that the data collection yields the dependent variable scores shown in Table 8.9.

Computing the values necessary for equation 8.1 we find $\Sigma D = 4$,

Table 8.9

Dependent Variable Scores

Level on Initial Measure	*Experimental Group*	*Scores Control Group*	*D*
1	10	5	5
2	1	8	−7
3	3	4	−1
4	5	1	4
5	7	6	1
6	2	4	−2
7	9	5	4

[6] s stands for standard deviation, and is discussed on p. 154.
[7] But not without some cost. For by discarding subjects we are possibly destroying the representativeness of our sample. Hence the confidence that we can place in our generalization to our population is reduced. We might also add that we would be interested in comparing the groups on the basis of a measure of variability of the scores. In this case the groups would be rather well matched with regard to their variability as evidenced by the standard deviations of 5.08 and 4.58 for the experimental and control groups respectively.

$\Sigma D^2 = 112$, $\overline{X}_E = 5.28$ and $\overline{X}_c = 4.71$.[8] Substituting these values in equation 8.1:

$$t = \frac{5.28 - 4.71}{\sqrt{\dfrac{112 - (4)^2/7}{(7)(6)}}} = 0.35$$

Entering our Table of t with a value of 0.35 and with 6 degrees of freedom we find that the t is not significant beyond the 5 per cent level. Hence we may conclude that the particular training program evaluated does not increase the efficiency of performance in a panic situation (for purposes other than illustration we would require a larger number of subjects to reach this conclusion). In retrospect we might compute the value of the correlation between the matching variable and the dependent variable scores, and find that it is too low (it would be in the neighborhood of .10) to have been worthwhile—we should have obtained a better matching criterion if we were to use a matched-groups design in this experiment.

Statistical Analysis with the A Test

The statistical test that we have presented for analyzing the two matched groups design has been the t test. We may use different statistical tests in analysing the results obtained from any particular experimental design. The t test has been used for the matched-groups design for several reasons: (1) because we used it for the randomized groups design, it was convenient to expand it for the present design; (2) because it is the test that is generally used; and (3) because we want to consider several important characteristics of experimentation by referring to the generalized formula for t as it applies to the matched-groups design. There is, however, a computationally simpler test that can be used for this design—the A test (Sandler, 1955). This statistic has been rigorously derived from the t ratio (equation 8.3); hence the two tests always yield the same results. The equation for computing A is:

(8.2) $$A = \Sigma D^2/(\Sigma D)^2$$

[8] Incidentally, we might make use of a principle of statistics in computing the numerator of the t test for the matched-groups design: that the difference between the means is equal to the mean of the differences of the paired observations. Therefore, as a short-cut, instead of computing the means of the two groups and subtracting them, as we have done, we could divide the sum of the differences (ΣD) by n and obtain the same answer: $\Sigma D/u = 4/7 = .57$.

Table 8.10

Table of A

(For any given value of $n - 1$, the table shows the values of A corresponding to various levels of probability. A is significant at a given level if it is equal to or *less than* the value shown in the table.)

Probability

$n - 1$	$0 \cdot 10$	$0 \cdot 05$	$0 \cdot 02$	$0 \cdot 01$	$0 \cdot 001$	$n - 1$
1	$0 \cdot 5125$	$0 \cdot 5031$	$0 \cdot 50049$	$0 \cdot 50012$	$0 \cdot 5000012$	1
2	$0 \cdot 412$	$0 \cdot 369$	$0 \cdot 347$	$0 \cdot 340$	$0 \cdot 334$	2
3	$0 \cdot 385$	$0 \cdot 324$	$0 \cdot 286$	$0 \cdot 272$	$0 \cdot 254$	3
4	$0 \cdot 376$	$0 \cdot 304$	$0 \cdot 257$	$0 \cdot 238$	$0 \cdot 211$	4
5	$0 \cdot 372$	$0 \cdot 293$	$0 \cdot 240$	$0 \cdot 218$	$0 \cdot 184$	5
6	$0 \cdot 370$	$0 \cdot 286$	$0 \cdot 230$	$0 \cdot 205$	$0 \cdot 167$	6
7	$0 \cdot 369$	$0 \cdot 281$	$0 \cdot 222$	$0 \cdot 196$	$0 \cdot 155$	7
8	$0 \cdot 368$	$0 \cdot 278$	$0 \cdot 217$	$0 \cdot 190$	$0 \cdot 146$	8
9	$0 \cdot 368$	$0 \cdot 276$	$0 \cdot 213$	$0 \cdot 185$	$0 \cdot 139$	9
10	$0 \cdot 368$	$0 \cdot 274$	$0 \cdot 210$	$0 \cdot 181$	$0 \cdot 134$	10
11	$0 \cdot 368$	$0 \cdot 273$	$0 \cdot 207$	$0 \cdot 178$	$0 \cdot 130$	11
12	$0 \cdot 368$	$0 \cdot 271$	$0 \cdot 205$	$0 \cdot 176$	$0 \cdot 126$	12
13	$0 \cdot 368$	$0 \cdot 270$	$0 \cdot 204$	$0 \cdot 174$	$0 \cdot 124$	13
14	$0 \cdot 368$	$0 \cdot 270$	$0 \cdot 202$	$0 \cdot 172$	$0 \cdot 121$	14
15	$0 \cdot 368$	$0 \cdot 269$	$0 \cdot 201$	$0 \cdot 170$	$0 \cdot 119$	15
16	$0 \cdot 368$	$0 \cdot 268$	$0 \cdot 200$	$0 \cdot 169$	$0 \cdot 117$	16
17	$0 \cdot 368$	$0 \cdot 268$	$0 \cdot 199$	$0 \cdot 168$	$0 \cdot 116$	17
18	$0 \cdot 368$	$0 \cdot 267$	$0 \cdot 198$	$0 \cdot 167$	$0 \cdot 114$	18
19	$0 \cdot 368$	$0 \cdot 267$	$0 \cdot 197$	$0 \cdot 166$	$0 \cdot 113$	19
20	$0 \cdot 368$	$0 \cdot 266$	$0 \cdot 197$	$0 \cdot 165$	$0 \cdot 112$	20
21	$0 \cdot 368$	$0 \cdot 266$	$0 \cdot 196$	$0 \cdot 165$	$0 \cdot 111$	21
22	$0 \cdot 368$	$0 \cdot 266$	$0 \cdot 196$	$0 \cdot 164$	$0 \cdot 110$	22
23	$0 \cdot 368$	$0 \cdot 266$	$0 \cdot 195$	$0 \cdot 163$	$0 \cdot 109$	23
24	$0 \cdot 368$	$0 \cdot 265$	$0 \cdot 195$	$0 \cdot 163$	$0 \cdot 108$	24
25	$0 \cdot 368$	$0 \cdot 265$	$0 \cdot 194$	$0 \cdot 162$	$0 \cdot 108$	25
26	$0 \cdot 368$	$0 \cdot 265$	$0 \cdot 194$	$0 \cdot 162$	$0 \cdot 107$	26
27	$0 \cdot 368$	$0 \cdot 265$	$0 \cdot 193$	$0 \cdot 161$	$0 \cdot 107$	27
28	$0 \cdot 368$	$0 \cdot 265$	$0 \cdot 193$	$0 \cdot 161$	$0 \cdot 106$	28
29	$0 \cdot 368$	$0 \cdot 264$	$0 \cdot 193$	$0 \cdot 161$	$0 \cdot 106$	29
30	$0 \cdot 368$	$0 \cdot 264$	$0 \cdot 193$	$0 \cdot 160$	$0 \cdot 105$	30
40	$0 \cdot 368$	$0 \cdot 263$	$0 \cdot 191$	$0 \cdot 158$	$0 \cdot 102$	40
60	$0 \cdot 369$	$0 \cdot 262$	$0 \cdot 189$	$0 \cdot 155$	$0 \cdot 099$	60
120	$0 \cdot 369$	$0 \cdot 261$	$0 \cdot 187$	$0 \cdot 153$	$0 \cdot 095$	120
∞	$0 \cdot 370$	$0 \cdot 260$	$0 \cdot 185$	$0 \cdot 151$	$0 \cdot 092$	∞

To illustrate the computation of A, we need merely refer to the data presented in Table 8.4. There we found that $\Sigma D = -13$, and $\Sigma D^2 = 39$. Hence $A = 39/(-13)^2 = 0.23$. We now need to determine the probability level for A. To do this we compute the degrees of freedom just as we did for equation 8.1, i.e., $df = n - 1$. Since $n = 5$ in this example, $df = 4$. And entering Table 8.10 with four degrees of freedom we read across the rows. We note that as we move to the right, the value of A, unlike the value of t in the t table, decreases. That is, the *smaller* the value of A, the smaller the probability associated with it. In the t table the *larger* the value of t, the smaller the associated value of P. Hence, we find that with $df = 4$, a value of $A = 0.23$ has a P of less than 0.01. And this is precisely what the results of the t test told us. For further practice with the A test, you might check the results of the second experiment Table 8.9 that we used as an example. (You should find that $A = 7.00$ and thus $P > 0.05$). The advantages of the A test over the t test for the two-matched-groups design should be apparent. It is to be expected that within the foreseeable future experimenters will increase the frequency with which they use the A test in preference to the t test for the present design.

Which Design to Use—Randomized-Groups or Matched-Groups?

One advantage of the matched groups design over the randomized groups design is that it assures approximate equality of the two groups. That equality is not helpful to us, however, unless it is equality as far as measures of the dependent variable are concerned. Hence, if the matching variable is highly correlated with the dependent variable, then the equality of groups is beneficial. If not, then it is not beneficial—in fact, it can be detrimental. If a high correlation obtains, we should prefer a matched-groups design. But how high is high? We shall now offer some guiding considerations to help answer this question. To do this let us point out a general disadvantage of the matching design. We have said that the formula for computing degrees of freedom is $n - 1$. The formula for degrees of freedom with the randomized-groups design is $N - 2$. In other words when using the matched-groups design you have fewer degrees of freedom available than with the randomized-groups design. In the example above, for instance, there were seven subjects in each group. Hence $n = 7$, or $N = 14$. With the matched-groups design we had $7 - 1 = 6$ degrees of freedom whereas if we had used the randomized-groups design instead, we would have had $14 - 2 = 12$. And we may recall that the greater the number of degrees of freedom available, the smaller the value of t required for significance. For this reason the matched-groups design suffers a disadvantage compared to the randomized-groups design.

It may happen that a given t would have been significant with the randomized-groups design but not with the matched-groups design. Suppose we obtained the same value of t regardless of the design used. The t, we might say, is 2.05 obtained with 16 subjects per group. With a matched-groups design we would have 15 df and find that a t of 2.131 is required for significance at the 5 per cent level—hence the t is not significant. With a randomized-groups design, however, it would be significant, for with the 30 df available we would need a value of only 2.042 for significance at the 5 per cent level.

To summarize the situation concerning the choice of a matched-groups or a randomized-groups design, the advantage of the former is that the value of t may be increased if there is a positive correlation between the matching variable and the dependent variable. On the other hand one loses degrees of freedom when using the matched-groups design—half as many degrees of freedom are available with it as with the randomized-groups design. Therefore, if the correlation is going to be large enough to more than offset the loss of degrees of freedom, then one should use the matched-groups design.[9] If it is not, then the randomized-groups design should be used.[10] In short, if one is to use the matched groups design he should be rather sure that the correlation between his matching and his dependent variable is rather high and positive.

At this point a bright student might say: "Look here: you have made so much about this correlation between the matching and the dependent variable and I understand the problem. You say to try to find some previous evidence that a high correlation exists. But maybe this correlation doesn't hold up in your own experiment. I think I've got this thing licked. Let's match our subjects on what we think is a good variable and then actually compute the correlation. If we find that the correlation is not sufficiently high, then let's forget that we matched subjects and simply run a t test for a randomized groups design. If we do this, we can't lose—either the correlation is pretty high and we offset our loss of degrees of freedom using the matched-groups design or it is too low so we use a randomized-groups design and don't lose our degrees of freedom."

"This student," we might say, "is thinking, and that's good. But what he's thinking is wrong." An extended discussion of what is wrong with the thinking must be left to a course in statistics, but we can say that

[9] It might be observed that if the number of subjects in a group is large (e.g., if $n > 30$), then one can afford to lose degrees of freedom by matching. That is, there is such a small difference between the value of t required for significance at any given level with a large df that one would not lose much by matching subjects even if the correlation between the independent and dependent variables is zero. Hence the loss of df consideration is only an argument against the matched groups design when n is relatively small.
[10] An elaboration of these statements is offered in the Appendix.

the error is similar to that previously referred to in setting the level of significance for t (p. 91). There we said that the experimenter may set whatever level of significance he desires, providing he does it before he conducts his experiment. Analogously, the experimenter may select whatever design he wishes, providing he does it before he conducts his experiment. And in either case the decision must be adhered to. For where he chooses a matched-groups design he has also mortgaged himself to a certain type of statistical test (e.g., the matched t test, which has a certain probability attached to its results). And if he changes the design he disturbs the probability that he can assign to his t through the use of the t Table. Hence, if an experimenter decides to use a matched-groups design, he must stick to that decision. Perhaps the following experience might be consoling to you in case you ever find yourself in the unlikely situation described. The author once helped conduct an experiment in which a matched-groups design was used (McGuigan & Mac-Caslin, 1955a). Previous research had shown that the correlation between the variable that was used to match subjects and the dependent variable was 0.72. This was an excellent opportunity to use a matched-groups design. However it turned out that the correlation was −0.24 for the data collected. And we shall see what a negative correlation does to the value of t. Hence in the author's experiment not only were degrees of freedom lost, but as we shall see (the value of the statistical test that would correspond to) the value of t was actually decreased.

Appendix to Chapter 8

It is particularly advantageous to consider several additional matters that should facilitate an understanding of the two-matched-groups design. To do this we shall consider the *generalized* equation for the t test—*generalized* in that it is applicable to either the randomized groups or the matched-groups design. It may be written:[11]

$$(8.3) \qquad t = \frac{\overline{X}_1 - \overline{X}_2}{\sqrt{\dfrac{s_1^2}{n_1} + \dfrac{s_2^2}{n_2} - 2(r_{12})\left(\dfrac{s_1}{\sqrt{n_1}}\right)\left(\dfrac{s_2}{\sqrt{n_2}}\right)}}$$

[11] It is important not to forget that sample characteristics (statistics) are used as estimates of corresponding population characteristics (parameters). The (population) parameters are fixed, but unknown, while the (sample) statistics can be expected to vary from sample to sample. Since our emphasis is on how to compute statistics, we are using notation for sample statistics in our discussions. Your further work will lead to a greater appreciation of the distinction, which can be more clearly made by contrasting the notation for statistics and parameters, and such as $\overline{X}$ and μ for the mean, s and σ for the standard deviation, and r and ρ for correlation.

The two previous equations for t (equations 5.2 and 8.1) are derivatives of equation 8.3. To understand equation 8.3 we need to understand the following new symbols: s (the standard deviation) and s^2 (the variance), both of which are measures of variability, and r_{12}.

The Standard Deviation and Variance

Suppose that we are interested in making some statements about the intelligence of the students at a given college. There may be 1000 students in the college; this would give us 1000 scores with which to deal, a very cumbersome number. If someone asked us about the intelligence of the students in the college we might start reading the scores to him. Before we could reach the thousandth score, however, our inquirer undoubtedly would have withdrawn his question. A much more reasonable procedure for telling him about the intelligence scores of our college students would be to resort to certain statistics for the purpose of reducing the data, thus allowing us to make certain summary statements about them. We could, for instance, compute a mean and tell our inquirer that the mean intelligence of the student body is 125, or whatever. While this would be accurate, it would not be adequate, for there is more to the story than that. Whenever we seek to describe a group of data we need to offer two kinds of statistics—*a measure of central tendency and a measure of variability*. Measures of central tendency tell us something about the central point of value of a group of data. They are kinds of averages that tell us what the typical score in a distribution of data is. The most common measure of central tendency is the mean. Measures of variability, tell us how the scores are spread out—they indicate something about the nature of the distribution of scores. In addition to telling us this, they also tell us about the range of scores in the group. The most frequently used measure of variability, probably because it is usually the most reliable of these measures (in the sense that it varies least from sample to sample), is the standard deviation. The standard deviation is usually symbolized by s.

To illustrate the importance of measures of variability we might imagine that our inquirer says to us: "Fine. You have told me the mean intelligence of your student body, but how homogeneous are your students? Do their scores tend to concentrate around the mean, or are there many that are considerably above the mean and many considerably below the mean?" To answer this, we might resort to the computation of the standard deviation. The larger the standard deviation, the more variable are our scores. To illustrate, let us assume that we have collected the intelligence scores of students at two different colleges. Plotting the

number of people who obtained each score at each college we might
obtain the distributions shown in Figure 8.1.

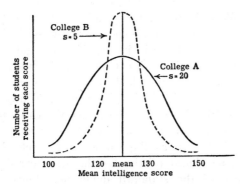

Figure 8.1. Distribution of intelligence scores at two colleges.

By computing the standard deviation[12] for the two groups, we might
find their values to be 20 for College A and 5 for College B. Comparing
the distributions for the two colleges, we note that there is considerably
more variability in College A than in College B. That is, the scores for
College A are more spread out or scattered than for College B. And
this is precisely what our standard deviation tells us—the larger the
value for the standard deviation, the greater the variability of the
distribution of scores. The standard deviation (for a normal distribution) also gives us the more precise bit of information that about
two-thirds of the scores fall within one standard deviation of the mean.
To illustrate, let us first note that the mean intelligence of the two
colleges is the same, 125. If we subtract one standard deviation (i.e.,
20) from the mean for College A and add one standard deviation
to that mean, we obtain two values: 105 ($125 - 20 = 105$) and 145
($125 + 20 = 145$). Therefore about two-thirds of the students in College

[12] You are advised that the statistics discussed in the chapter are, in the main, concerned only with values for samples. From the sample values the population values
may be inferred. This is another case where we must limit our consideration of
statistical matters to those that are immediately relevant to the conduct of experiments. But you are again advised to pursue these important topics by further
work in statistics. An equation for computing the standard deviation for a sample is
$s = \sqrt{\dfrac{n\Sigma X^2 - (\Sigma X)^2}{n(n-1)}}$ As Guilford (1956) points out, the estimate of the population
value for the standard deviation would be slightly different but "when N is large
(30 or greater) . . . "the two [computed] values are practically identical." You
should be able to compute s on the basis of the knowledge that you now possess.
To check yourself you might compute s for the dependent variable scores of the
two groups in Table 8.3. You should find that for the Reading Group $s = 2.61$ and
for the Reading and Reciting Group $s = 2.17$.

A have an intelligence score between 105 and 145. Similarly, about two-thirds of the students at College B have scores between 120 (125 − 5) and 130 (125 + 5). Hence, we have a further illustration that the scores at College A are more spread out than at College B. We might for a moment speculate about these student bodies. College A, we might guess, is rather loose in its selection of students, as might be the case in some state universities. College B is much more selective, having a rather homogeneous student body. Such a sample might occur for a private institution with high tuition costs. In any event we wish to make only one point here, that the larger the value of the standard deviation, the more variable (spread out) the scores.

The symbol s^2 is known as the *variance* of a set of scores. It has essentially the same characteristics as the standard deviation and is merely the square of the standard deviation. Hence, if $s = 5$, then $s^2 = 25$. To illustrate these statistics further, let us assume that we have obtained the dependent variable scores for the two groups in an experiment shown in Table 8.11.

Table 8.11

Some Dependent Variables Scores for Two Groups of Subjects

Experimental Group		Control Group	
Subject	Score	Subject	Score
1	10	1	7
2	1	2	6
3	0	3	7
4	5	4	5
5	3	5	6
6	7	6	7
7	9	7	6
8	6	8	5
9	8	9	7
10	2	10	7

The scores for the experimental group vary from 0 to 10; the standard deviation here is 3.48. The scores for the control group, on the other hand, are much less variable. In fact, all the scores are either 5, 6, or 7. We should expect that the standard deviation for the control group is considerably smaller than for the experimental group. We shall not be disappointed, for its computation yields a value of .82. In turn, the variance is 12.10 for the experimental group and .68 for the control group. Using the variances as indices of variability, we can see that they also show that the variability of the experimental group is greater than that of the control group. Incidentally, we might note that if all the scores for one group were the same, say 7, both the standard deviation and the variance would be zero, for there would be zero variability among the scores.

The Nature of r_{12}.

With this understanding of the standard deviation and variance, let us now turn to the symbol r_{12}, the last unfamiliar symbol in equation 8.3. We have already discussed the general nature of a correlation, but it remains for us to specify this particular correlation. It stands for the (linear) correlation between the dependent variable scores of the two groups of subjects in a matched-groups design. (Let us observe that r_{12} is read "the correlation between the dependent variable scores of group 1 and group 2" and not "r-twelve".) That is, in this type of design, we have paired subjects in one group with subjects in the other group. And these pairs are ranked on the basis of their matching variable scores. Thus, the highest pair of scores on the matching variable is ranked first, the second highest pair ranked second, and so on. If the dependent variable scores of the first pair of subjects are the highest scores in each group, if the second pair of subjects provided the second highest dependent variable scores in their respective groups, and so on down the rank of pairs without exception, *then* the correlation between these two sets of dependent variable scores would be perfect. That is, r_{12} would equal 1.0.[13] Similarly, if there are only a few exceptions in this order, r_{12} would be high, but less than perfect. And so on for the other possibilities, as discussed in the previous section on correlation.

To cement the nature of r_{12} refer back to Table 8.3 and note the sets of dependent variable scores for each of our two groups. If we correlated these two sets of scores, we would find that $r_{12} = 0.90$.[14] That is, the highest pair of subjects on the matching variable has the highest set of dependent variable scores in its group, the lowest pair of subjects has the lowest dependent variable scores, and so on.

The nature of r_{12} should now be clear. And you could compute it for any set of data that you want, simply by applying the equation in footnote 2 of this chapter (see p. 142). But what is its significance? To answer this question, let us restate a principle that was stressed earlier. For the matched-groups design to be successful you should have a reasonably high correlation between the matching variable and the dependent variable scores. Strictly speaking, this latter correlation does not enter into your statistical analysis, but is taken account of only

[13] Again on the assumption that the increase in scores for each group is proportional (see footnote 3, p. 143). If they are not, then the correlation will be somewhat less than 1.0.
[14] Again, emphasizing that we are considering only linear relationships, ignoring curvilinearity.

indirectly. That is, rather than using an actual value for the matching variable–dependent variable correlation in our t-test, we use the value of r_{12}. And this is possible because the value of r_{12} is an indication of the value of the matching variable–dependent variable correlation.[15] That is, if the matching variable–dependent variable correlation is high, r_{12} will be high; and if the matching variable-dependent variable correlation is low, r_{12} will also be low. The reasonableness of these statements should be apparent after a little reflection. The only reason that we would expect a high value of the correlation between the paired dependent variable scores of our two groups is that they were matched together on the basis of a variable that correlates with the dependent variable. Thus, the reason that the pair of subjects who were ranked first on the matching variable should both exhibit the highest scores on the dependent variable is that there is a correlation between the matching variable and the dependent variable. On the other hand, if the correlation between the matching and the dependent variable scores is zero, we would expect the value of r_{12} to be zero, for there would be no reason to expect that the top-ranked pair of subjects on the matching variable should both exhibit the highest dependent variable scores.

The situation at this point is that r_{12} is the correlation between the dependent variable scores of pairs of subjects in a matched-groups design. Since r_{12} is an indication of the value of the correlation between the matching and the dependent variable scores, we use it in conducting our statistical analysis.

Let us now take a broader look at equation 8.3. If we want to increase our chances of obtaining a significant t there are two courses of action that can be followed. First, we can attempt to increase the value of the numerator (the difference between the means of the two groups), or to decrease the value of the denominator. As the value of the numerator increases, or as the value of the denominator decreases, the value of t increases. And we know that the larger t is, the more likely it is to be significant. To illustrate, let us say that the numerator is 5 and the denominator is 10. In this case t is:

$$t = 5/10 = 0.50$$

But if we are able to decrease the value of the denominator to 2, with no change in the numerator, we would have:

$$t = 5/2 = 2.50$$

And a t of 2.50 is likely to be significant, whereas one of 0.50 is not.

[15] Here is a good example of what we said in footnote 11, i.e., while there is no exception in the rankings of the two sets of scores, the intervals between the scores in each set are not proportional, and hence the value of r only *approaches* 1.0.

Our question should now be how we can decrease the value of the denominator of equation 8.3. This matter is discussed more thoroughly in Chapter 14, but let us consider one possible way here. We may note that the larger the variance of the two groups, the larger is the denominator. If s_1^2 and s_2^2 are each 10, the denominator will be larger than if they are both 5. But we may note that from the variances we subtract r_{12} (and also s_1 and s_2, but these need not concern us here). And any subtraction from the variances of the two groups will result in a smaller denominator with, as we said, an attendant increase in t. Furthermore, we said, the size of r_{12} depends on the correlation between the matching variable and the dependent variable. Hence, if that correlation is large and positive, we may note that the denominator is decreased.

By way of illustration, assume that the difference between the means of the two groups is 5 and that there are 9 subjects in each group (n_1 and n_2 both equal 9). Further assume that s_1 and s_2 are both 3 (hence s_1^2 and s_2^2 are both 9) and that r_{12} is 0.70. Substituting these values in equation 8.3 we obtain:

$$t = \frac{5}{\sqrt{9/9 + 9/9 - 2(0.70)(3/\sqrt{9})(3/\sqrt{9})}} = 6.49$$

It should now be apparent that the larger the positive value of r_{12}, the larger is the term that is subtracted from the variances of the two groups. In an extreme case of the above illustration, where $r_{12} = 1.0$, we may note that we would subtract 2.00 from the sum of the variances (2.00); this leaves a denominator of zero, in which case t might be considered to be infinitely large. On the other hand, suppose that r_{12} is rather small—say it is 0.10. In this case we would merely subtract 0.20 from 2.00, and the denominator would be only slightly reduced. Or if $r_{12} = 0$, then it can be seen that zero would be subtracted from the variances, not reducing them at all. The moral should now be clear: *the larger the value of r_{12} (and hence the larger the value of the correlation between the matching variable and the dependent variable), the larger the value of t.*

Now let us consider the heart of the matter: How large should be the correlation between the matching and the dependent variable in order to prefer a matched-groups design to a randomized-groups design? Well, we can't answer this question, at least not in this form. But we can give a very good answer by changing it somewhat. This is, in fact, the principal reason why we have stressed the nature of r_{12} and its relationship with the correlation between the matching and the dependent variable. Thus, we shall answer the question: How large should r_{12} be before a matched-groups design should be preferred? And the answer is given

in Fig. 8.2.[16] To use Fig. 8.2 you merely select the number of subjects
in each group (n) and enter the figure at that value on the horizontal

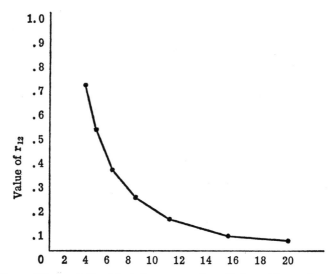

Figure 8.2. A relationship between n and r_{12}. Enter with a value of
n and read the value of your expected value of r_{12} that intersects the
curve at that point. If your expected value of r_{12} exceeds the
value obtained from the cure, a matched-groups design is to be preferred.

axis. Then read up until you reach the curve. The value of the curve at
that point on the vertical axis indicates the necessary minimum value
of r_{12} in order for a matched-groups design to be preferred. For instance,
if you want to know the minimal value of r_{12} that you will need in the
event that you have 12 subjects in each group, you would find "12" on
the horizontal axis. Then reading up to the curve and over to the vertical
axis you find that r_{12} is about 0.17. *In order to prefer a matched-groups
design,* then, you should have a minimal value of 0.17 for r_{12}.

One final consideration of the value of r_{12} is what the effect of a negative correlation would be on the value of t? A little reflection should
reveal that a negative correlation *increases* the denominator, thus decreasing t. In this case, instead of subtracting from the variances, we
would have to add to them ("a minus times a minus gives us a plus").
Furthermore, the larger the negative correlation, the larger our denominator becomes. For example, suppose that in the previous example instead

[16] We are simplifying the situation considerably by presenting a rather conservative
case. It can rather safely be said that if the value of your r_{12} is at least that indicated for a certain number of subjects in Fig. 8.2, then you should prefer a
matched-groups design.

of having a value of $r_{12} = 0.70$, we had $r_{12} = -0.70$. In this case we can see that our computed value of t would decrease from 6.49 to 2.72. That is,

$$t = \frac{5}{\sqrt{9/9 + 9/9 - 2(-0.70)(3/\sqrt{9})(3/\sqrt{9})}} = 2.72$$

We previously said that equation 8.3 is a generalized formula, applicable to either of the two designs that we have discussed. One might ask, however, in what way it is applicable to the randomized-groups design, for it contains a correlation term and we have not referred to any correlation when using it—it is absurd, for instance, to talk about the correlation between pairs of subjects on the dependent variable when using the randomized-groups design, for by its very nature subjects are not paired. The answer to this is that since subjects have not been paired the correlation between any random pairing of subjects in the long run is zero. That is, if we randomly selected any subject in an experimental group, paired him with a randomly selected subject in the control group, and continued this procedure for all subjects, we would expect the correlation between the dependent variable scores to be zero (or more precisely, the correlation would not be significantly different from zero)—there simply would be no reason to expect other than a zero correlation since the subjects were not paired together on more than a chance basis. When using the randomized-groups design, we assume that r_{12} of equation 8.3 is zero. And being zero, the term that includes r_{12} "drops out." Thus, equation 8.3 assumes the following form for the randomized-groups design:

$$(8.4) \qquad t = \frac{\overline{X}_1 - \overline{X}_2}{\sqrt{s_1^2/n_1 + s_2^2/n_2}}$$

One final note. We have labeled the type of design discussed in this chapter as *the matched-groups design* where we have limited our discussion to the case of two groups. The two groups may be said to be matched because we paired subjects with similar scores. Since all subjects were paired together, the groups had to be approximately equivalent. This fact may be determined by comparing the distribution of matching scores for the two groups. The best such comparison would probably be to compare the means and standard deviations of the two groups. We would expect to find that the two groups would be quite similar on these two measures. However, the same result could also be achieved in other ways. That is, two groups could be formed in other ways so that they would have similar means and standard deviations. For example, we could simply assign subjects to two groups so that their total scores would be similar—no subjects would be paired together, but the means and standard deviations of the two groups would be approximately the same. Therefore, it may be considered that the technique of pairing subjects together is a specific type of design that results in matched groups. For this reason it could as well be called *the paired-groups design* to distinguish it from alternative procedures that result in matched groups. Alternative procedures, however, require different statistical analyses than that presented here (McNemar, 1955). Since they are not so generally used, nor judged to be as effective, they will not be considered here.

The two-matched-groups design (or if you prefer, the two-paired-groups design) implies that the design could be extended to more than two groups. For a discussion of a matched-groups design for more than two groups you are referred to Edwards (1950).

Summary of the Computation of t for a Two-Matched-Groups Design

Assume that two groups of subjects have been matched on an initial measure as indicated, and that the following dependent variable scores have been obtained for them.

Initial Measure	Group 1	Group 2
1	10	11
2	10	8
3	8	6
4	7	7
5	7	6
6	6	5
7	4	3

1. The equation for computing t (equation 8.1) is:

$$t = \frac{\overline{X}_1 - \overline{X}_2}{\sqrt{\dfrac{\Sigma D^2 - (\Sigma D)^2/n}{n(n-1)}}}$$

2. Compute the value of D for each pair of subjects, and then the sum of D (ΣD), the sum of the squares of D (ΣD^2), the sum of D squared $[(\Sigma D)^2]$, and n.

Initial Measure	Group 1	Group 2	D
1	10	11	−1
2	10	8	2
3	8	6	2
4	7	7	0
5	7	6	1
6	6	5	1
7	4	3	1

$$\Sigma D = 6$$
$$\Sigma D^2 = 12$$
$$n = 7$$

3. Determine the difference between the means. This may be done by computing the mean for each group and subtracting them, or by computing the mean of the differences. Since the latter is easier, we shall do this.

$$\text{Mean of the differences} = \Sigma D/n = 6/7 = 0.86$$

4. Substitute the above values in equation 8.1:

$$t = \frac{0.86}{\sqrt{\dfrac{12 - (6)^2/7}{7(7-1)}}}$$

5. Perform the operations as indicated and determine the value of t.

$$t = \frac{0.86}{\sqrt{0.16}} = 2.15$$

6. Determine the number of degrees of freedom associated with the computed value of t.

$$df = n - 1 = 7 - 1 = 6$$

7. Enter the table of t with the computed values of t and df. Determine the probability associated with this value of t. In this example, $0.1 > P > 0.05$. Therefore, assuming a significance level of 0.05, the null hypothesis is not rejected.

Problems

1. A psychologist seeks to test the hypothesis that the western grip for holding a tennis racket is superior to the eastern grip. He matches his subjects on the basis of a physical fitness test, trains them in the use of these two grips, respectively, and obtains the following scores on their tennis-playing proficiency. Assuming adequate controls, that a 0.05 level of significance is set, and that the higher the score the better the performance, what can he conclude with respect to his empirical hypothesis?

Rank on Matching Variable	Score on Dependent Variable	
	Eastern Grip Group	*Western Grip Group*
1	10	2
2	5	8
3	9	3
4	5	1
5	0	3
6	8	1
7	7	0
8	9	1

2. To test the hypothesis that the higher the induced anxiety, the

better the learning, an experimenter formed two groups of subjects by matching them on an initial measure of anxiety. He then induced considerable anxiety into his experimental group, but not into his control group. He then obtained the following scores on a learning task, where the higher the score, the better the learning. Assuming adequate controls were exercised and that he set a significance level of 0.05, did he confirm his hypothesis?

Rank on Matching Variable	Dependent Variable Scores	
	Experimental Group	*Control Group*
1	8	6
2	8	7
3	7	4
4	6	5
5	5	3
6	3	1
7	1	2

3. A military psychologist wishes to evaluate a training aid that was designed to facilitate the teaching of soldiers to read a map. He forms two groups of subjects, matching them on the basis of a visual perception test (an ability that is important in the reading of maps). He sets a significance level of 0.02 and exercises proper controls. Assuming that the higher the score, the better the performance, did the training aid facilitate map reading proficiency?

Rank on Matching Variable	*Scores of Group That Used the Training Aid*	*Scores of Group That Did Not Use the Training Aid*
1	30	24
2	30	28
3	28	26
4	29	30
5	26	20
6	22	19
7	25	22
8	20	19
9	18	14
10	16	12
11	15	13
12	14	10
13	14	11
14	13	13
15	10	6
16	10	7
17	9	5
18	9	9
19	10	6
20	8	3

9

Experimental Design: The Case of More Than Two Randomized Groups

Concerning the Value of Using More Than Two Groups

Among the more frequently used designs in psychological research are those in which more than two groups is used. Suppose a psychologist had two methods of remedial reading available. The methods are both presumably helpful to students who have not adequately learned to read by the usual method, but he wishes to know which method is superior. Furthermore, he wants to know whether either of these methods is actually superior to the normal method for such problem cases. To answer these questions, he might design an experiment which involves three groups of subjects.

If he has available sixty students who show marked reading deficiencies, his first step would be randomly to assign them to three groups. Assume that he assigns an equal number of subjects to each group, although this need not be the case. The first group would be taught to read by using Method A, and the second group by Method B. A comparison of the results from these two groups would tell him which, if either, is the superior method. He also wants to know if either method is superior to the normal method of teaching, which has heretofore been ineffective with this group. So, he would have his third group continue training under the normal method, as a control group. After a certain period of time, perhaps nine months, a standard reading test might be administered to the three groups. A comparison of the reading proficiency of the three groups on this test should answer the questions.

It is also possible to answer these questions by conducting a series of separate two groups experiments. It would be possible, for instance, to conduct one experiment in which Method A is compared to Method B, a second in which Method A is compared to the control condition, and a third in which Method B is compared to the control condition. Such a procedure is obviously less desirable, for not only would more work be required but the problem of controlling extraneous variables would be sizeable. For example, we would wish to hold the experimenter variable constant, so the same experimenter should conduct all three experiments. Even so, it is likely that the experimenter would behave differently in the first and last experiments, perhaps due to improvement in his teaching proficiency, or even the production of boredom or fatigue. Therefore, the design in which three groups are used simultaneously is superior in that less work is required, fewer subjects are used, and experimental control is better.

The randomized-groups design for the case of more than two groups may be applied to a wide variety of problems. To illustrate we might list some problems suggested by Edwards that are amenable to this type of design: ". . . the influence of different periods of food deprivation upon learning; the influence of different numbers of reinforcements upon conditioning and extinction; the influence of different methods of instructions upon achievement; the influence of different sets of verbal instructions upon problem solving; the influence of different sensory cues upon maze learning; the influence of different kinds of motivation upon performance; the influence of different periods of rest upon fatigue; the influence of different kinds of interpolated activities upon learning; the influence of different kinds of work situations upon production and fatigue; the influence of different periods of practice upon learning" (Edwards, 1950).

The procedure for applying a multi-group design (i.e., a design with more than two groups) to any of the above problems would be to select several values of the independent variable and assign a group of subjects to each value. For example, to study the influence of different periods of food deprivation upon performance, we might choose the following values of the independent variable: 0 hours, 1 hour, 12 hours, 24 hours, 36 hours, and 48 hours of deprivation. Having selected six values of the independent variable we would have six different groups of subjects, probably animals. To study the influence of different periods of practice upon learning we might select four values of the independent variable: 0, 5, 10, and 15 hours. We would then randomly assign our subjects to four groups and train one group under each condition.

It would be possible to study any of the above problems by conducting a single experiment using a two-groups design. In that event we would select only two values of the independent variable. A sizeable advantage of a multi-groups design becomes apparent when one tries to decide which two values of the independent variables to use. Consider the above example concerning the influence of different periods of food deprivation upon performance. We previously selected six values of the independent variable to study. Which two would we use for a two-groups design? It would be advisable to have a control condition, so we would probably choose a zero hour period for one group. The second group might be trained under a 24-hour period.

Now, let us imagine that the six-groups design yields the following results: no difference in performance among the zero, 1, 12, and 24 hour conditions, but the 36 and 48 hour conditions are superior to the first three. The conclusion from this six-group experiment would be that variation of the length of food deprivation from zero to 24 hours does not affect performance, but greater periods of deprivation (specifically periods of 36 and 48 hours) increases performance. But if the two-groups design (using only zero and 24 hour deprivation periods) were applied to this problem, the results would *suggest* that variation of the length of deprivation does not affect performance, a conclusion that would be in error. Thus, in general, it should be apparent that the more values of the independent variable sampled, the better our evaluation of its influence on a given dependent variable.

Research in any given area usually progresses through two stages: first, we seek to determine which of many possible independent variables influences a given dependent variable; and second, when a certain independent variable has been identified as influential on a dependent variable, we attempt to establish the precise relationship between them. While it is possible that a two-groups design would accomplish the first purpose, it could not accomplish the second. For an adequate relationship cannot usually be established with only two values of the independent variable (and therefore also only two values of the dependent variable). To illustrate this point refer to Figure 9.1, where the values of an independent variable are indicated on the horizontal axis while the cor-

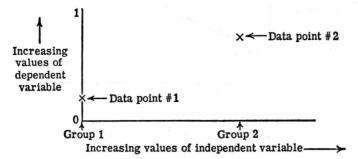

Figure 9.1. Two data-points obtained from a two-groups design. Group 1 was given a zero value of the independent variable while Group 2 was given a positive value. The value of the dependent variable is less for Group 1 (data point #1) than for Group 2 (data point #2).

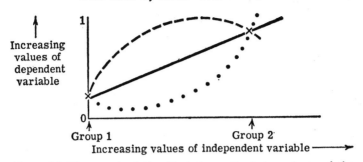

Figure 9.2. The actual relationship between the independent and the dependent variable is partially determined by the two data points. However, the curves that may pass through the two points are infinite in number. Three possible relationships are shown.

responding values of the dependent variable are read on the vertical axis.[1] The two plotted points (obtained from a two-groups design) indicate that as the value of the independent variable increases, the mean value of the dependent variable also increases. But this is a crude picture, for it tells us nothing about what happens between the two plotted points. See Figure 9.2 to illustrate a few of the infinite number of possibilities.

By using a three-groups design the relationship may be established more precisely. Let us say that we have the same two groups as in Figure 9.1, but in addition we have a third group that received a value of the independent variable halfway between those of the other two groups. Assuming that the mean dependent variable value for group 3 is that depicted in Figure 9.3, we would conclude that the relationship is prob-

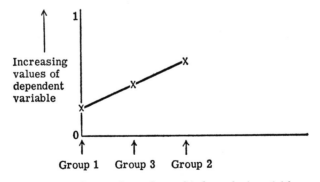

Figure 9.3. The addition of a third data point (Group 3) suggests that the relationship is a straight line function.

[1] The range of the independent variable in the following discussions should be clear from the context, e.g., from zero to infinity. We shall also assume that the data points are highly reliable and thus not the product of chance variation.

ably a straight-line function. Of course, we might be wrong. That is, the relationship is not necessarily the straight line indicated in Figure 9.3, for it is possible that some other relationship is actually the "true" one such as one of those shown in Figure 9.4. Nevertheless, with only three data points we prefer to bet that the straight line is the "true" relationship because it is the simplest of the several possible relationships. And experience suggests that it is reasonable to assume that the simplest curve gives the best predictions. That is, we would predict that if we obtain a data point for a new value of the independent variable (in addition to the three already indicated in Figure 9.3) that new data point would fall on the straight line. Different predictions would be made from the curves of Figure 9.4.

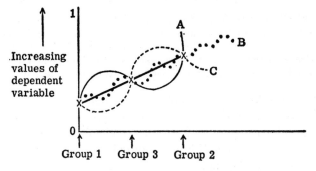

Figure 9.4. Other possible curves may pass through the three data points.

To illustrate, suppose that we add a fourth group whose independent variable value is halfway between those of groups 1 and 3. On the basis of the four relationships depicted in Figure 9.4 we could make four different predictions about the dependent variable value of this fourth group. First, using the straight-line function, we would predict that the data point for the fourth group would be that indicated below in Figure 9.5, i.e., if the straight line is the "true" relationship, the data point for the fourth group should fall on that line. The three curves of Figure 9.4, however, lead to three additional (and different) predictions.

Assume that when the data for the fourth group are analyzed, they indicate that the mean score is actually that indicated by the X_1 of Figure 9.5. This increases our confidence in the straight-line function—it, rather than the other possible functions, is probably the "true" one. If these were actually the results, our procedure of preferring the simplest curve as the "true" one (at least until contrary results are obtained) is justified. This procedure is what Reichenbach (1938) calls *inductive*

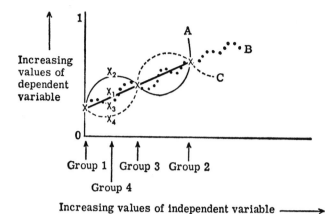

Increasing values of independent variable ⟶

Figure 9.5. Four predictions of a data point for Group 4. From the straight line of Figure 9.3 we would predict that the dependent variable score would be indicated by X_1. From the three curves of Figure 9.4 (curves A, B, and C) we would predict that the data point would be that indicated by the X_1, the X_2, and the X_3, respectively.

simplicity, the selection of the simplest curve that fits the data points. The safest induction is that the simplest curve provides the best prediction of additional data points. With the randomized design for more than two groups you can establish as many points as you like, consistent with the effort you can expend.

One general principle of experimentation when using a two-groups design is that it is advisable to choose rather extreme values of the independent variable.[2] If we had followed this principle, we would not have erred in the example concerning the influence of the period of deprivation upon performance. For instead of choosing zero- and 24-hour periods, as we did, we perhaps should have selected zero- and 48-hour periods. In this event the two-groups design would have led to a conclusion more like that of the six-groups design. However, it should still be apparent that the six-groups design yielded considerably more information, allowing us to establish the relationship between the two variables with a high degree of confidence. Even so, the selection of extreme values for two groups can lead to difficulties in addition to those already considered. To illustrate, assume that two data points are obtained, such as

[2] Let us emphasize the word "rather," for seldom would we ever want to select independent variable values for two groups that are *really* extreme. This is so because it is likely that all generalizations in psychology break down when the independent variable values are unrealistically extreme. Weber's law, which you probably studied in introductory psychology, is a good example. For while Weber's law holds rather well for weights that you can conveniently lift, it would obviously be absurd to state that it is true for extreme values of weights such as those of atomic size or those of several tons.

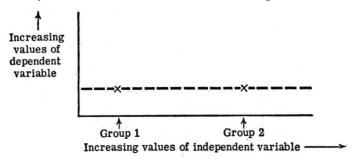

Figure 9.6. Two data-points for extreme values of the independent variable using a two groups design. These points suggest that the independent variable does not affect the dependent variable.

those indicated by the X's in Figure 9.6. Our conclusion would probably be that manipulation of the independent variable does not influence the dependent variable, for the dependent variable values for the two groups are the same. The best guess is that there is a lack of relationship as indicated by the horizontal straight line fitted to the two points.[3] Yet the actual relationship may be that indicated in Figure 9.7, a relationship that probably would have been uncovered by the use of a three-groups design. The corresponding principle with a three-groups design would be to select two extreme values of the independent variable, and also one value midway between them. Of course if the data point for Group 3 had been the same value as for Groups 1 and 2, then we would be more confident that the independent variable did not affect the dependent variable.

To summarize what we have said we may note that psychologists seek

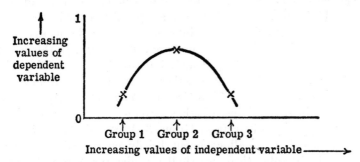

Figure 9.7. Postulated actual relationship for the data points of Figure 9.6. This relationship would be uncovered by a suitable three-groups design.

---

[3] One might quibble and say that the horizontal line is still a relationship. However, from our point of view this "relationship," if true, would indicate that variation of the independent variable does not differentially affect the dependent variable.

to determine which of a number of independent variables influence a given dependent variable, and also attempt to establish the relationship between them. With a two-groups design one is never sufficiently sure that he has selected the appropriate values of the independent variable in his attempt to determine whether or not that variable is effective. By using more than two groups, however, he increases his chances of (1) accurately determining whether a given independent variable is effective and (2) specifying the relationship between the independent and the dependent variable.

Statistical Analysis of a Randomized-Groups Design with More Than Two Groups

As in previous designs, we need to determine whether our groups differ significantly. However, we now have several groups to compare. As before we shall compare groups by comparing differences among their means. But what statistical procedure is most appropriate for this type of problem? Unfortunately for our present purposes, there is much disagreement among statisticians and among psychologists as to the correct answer to this question. In part, though only in part, the disagreements stem from different types of hypotheses that are being tested, and different aspects of the question that are emphasized. We here wish to minimize the extent to which we enter the various controversies, and to this end shall say that we are primarily interested in doing the following: (1) in making comparisons only between individual groups, that is, we are not interested in combining two or more groups to test these combined groups against some other group or combination of groups; (2) in making comparisons between all possible combinations of the separate groups. To make these points more concrete, let us say that you conduct an experiment in which there are three groups. In this event, we are saying that you would be primarily interested in determining whether Group 1 differs from Group 2, whether Group 1 differs from Group 3, and whether Group 2 differs from Group 3. You would not, for example, be interested in determining whether Groups 1 and 2, considered together as one group, differ from Group 3. With this limitation of our interest (although we shall comment briefly on procedures where our interest might be otherwise), we shall consider two types of statistical analysis for the more-than-two-randomized-groups design. After prolonged investigation of the numerous statistical procedures available, it was the author's opinion that Duncan's Range Test is the most appropriate. However this test is relatively new; the more widely used procedure by psychologists starts with the analysis of variance. By presenting both of these procedures (with the author's recommendation of Duncan's Range Test) you can make your own decision as to which you prefer to apply. It might be added, incidentally, that Duncan's Range Test is easier to apply, and is less time consuming.

Duncan's Range Test: A Three-Groups Design

To apply Duncan's Range Test, let us return to the remedial reading example where we had three groups: one trained under Method A, one under Method B, and the third received normal training. Assume that the scores in Table 9.1 (the higher the score, the better the performance) were obtained by the three groups on the reading test (the dependent variable):

Table 9.1

Fictitious Scores of Three Groups in an Experiment
on Remedial Reading Methods

Group A		Group B		Control-Group	
15	16	22	12	17	2
17	12	25	19	6	8
12	11	23	24	9	9
13	19	17	26	11	14
10	19	29	29	4	7
19	14	26	30	3	8
17	15	25	26	8	6
21	17	24	27	9	5
14	13	27	21	12	16
15	12	31	23	6	9

$$n = 20 \qquad\qquad n = 20 \qquad\qquad n = 20$$
$$\Sigma X = 301 \qquad\quad \Sigma X = 486 \qquad\quad \Sigma X = 169$$
$$\Sigma X^2 = 4{,}705 \qquad \Sigma X^2 = 12{,}208 \qquad \Sigma X^2 = 1{,}733$$
$$\bar{X} = 15.05 \qquad\quad \bar{X} = 24.30 \qquad\quad \bar{X} = 8.45$$

We seek to test for significant differences among the three groups or, more precisely, among the means of the three groups. The first step is to compute the sum of squares (SS) of the dependent variable scores for each group. The equation for the sum of squares of any group is

(9.1) $$SS = \Sigma X^2 - (\Sigma X)^2/n$$

Hence it can be seen that we need ΣX^2, ΣX, and n for each group. These values are computed as before and are presented in Table 9.1. We merely need to substitute them separately for each group in equation 9.1. For Group A the SS is:

$$SS_A = 4705 - (301)^2/20 = 174.95$$

The SS of Groups B and C (control group) can be seen to be:

$$SS_B = 12,208 - (486)^2/20 = 398.20$$
$$SS_C = 1,733 - (169)^2/20 = 304.95$$

Next we compute the square root of the error variance (s_e) (see p. 287) which for three groups, is given by the equation:

(9.2) $$s_e = \sqrt{\frac{SS_A + SS_B + SS_C}{3(n-1)}}$$

Where n is still the number of subjects in each group. Substituting the sum of squares for the three groups and n in equation 9.2 we find that s_e is:

$$s_e = \sqrt{\frac{174.95 + 398.20 + 304.95}{3(20-1)}} = 3.92$$

We now determine the degrees of freedom for s_e appropriate for Duncan's Range Test which is given by the equation:

(9.3) $$df = N - r$$

Where N is the total number of subjects in the experiment and r is the number of groups. This is found to be

$$df = 60 - 3 = 57$$

Assuming that we have set a 5% level of significance for testing the difference between two means we now need to refer to Table 9.2.[4] There we see that the columns are labeled "Number of Groups" and the rows "df." The values in the table are the "least significant standardized ranges" which are symbolized r_p.[5] We shall require various values of r_p for each test between two means that we shall make. Since we have three means in the present example we shall make the following tests: between the extreme means of three groups; between the highest and the middle means; and, between the lowest and the middle means. The p that is the subscript to the r indicates these three situations. For the first $p = 3$ and hence $r_p = r_3$, and for the other two $p = 2$ and hence, $r_p = r_2$. These symbols may be written r_3 and r_2. Therefore, we need to enter Table 9.2 to obtain the value of r_p when the extreme value of a group of three means is being tested, which is the column labeled "3." Finding the

---

[4] See Duncan (1955) for an elaboration of the precise nature of this probability value.
[5] Do not confuse r_p with r which is the number of groups. It is unfortunate that we cannot take the time to lay a broad base for our techniques of statistical analysis and for discussing these concepts in greater detail. The interested reader can of course always correct this deficiency by referring to other sources. For an elaboration of the concepts in this chapter you might refer to Duncan (1955) or to the textbook by Li (1957).

Table 9.2

Values of r_p for Duncan's Range Test (Significance Level = 5 Per cent)

df	Number of Groups															
	2	3	4	5	6	7	8	9	10	12	14	16	18	20	50	100
1	18.0	18.0	18.0	18.0	18.0	18.0	18.0	18.0	18.0	18.0	18.0	18.0	18.0	18.0	18.0	18.0
2	6.09	6.09	6.09	6.09	6.09	6.09	6.09	6.09	6.09	6.09	6.09	6.09	6.09	6.09	6.09	6.09
3	4.50	4.50	4.50	4.50	4.50	4.50	4.50	4.50	4.50	4.50	4.50	4.50	4.50	4.50	4.50	4.50
4	3.93	4.01	4.02	4.02	4.02	4.02	4.02	4.02	4.02	4.02	4.02	4.02	4.02	4.02	4.02	4.02
5	3.64	3.74	3.79	3.83	3.83	3.83	3.83	3.83	3.83	3.83	3.83	3.83	3.83	3.83	3.83	3.83
6	3.46	3.58	3.64	3.68	3.68	3.68	3.68	3.68	3.68	3.68	3.68	3.68	3.68	3.68	3.68	3.68
7	3.35	3.47	3.54	3.58	3.60	3.61	3.61	3.61	3.61	3.61	3.61	3.61	3.61	3.61	3.61	3.61
8	3.26	3.39	3.47	3.52	3.55	3.56	3.56	3.56	3.56	3.56	3.56	3.56	3.56	3.56	3.56	3.56
9	3.20	3.34	3.41	3.47	3.50	3.52	3.52	3.52	3.52	3.52	3.52	3.52	3.52	3.52	3.52	3.52
10	3.15	3.30	3.37	3.43	3.46	3.47	3.47	3.47	3.47	3.47	3.47	3.47	3.47	3.48	3.48	3.48
11	3.11	3.27	3.35	3.39	3.43	3.44	3.45	3.46	3.46	3.46	3.46	3.46	3.47	3.48	3.48	3.48
12	3.08	3.23	3.33	3.36	3.40	3.42	3.44	3.44	3.46	3.46	3.46	3.46	3.47	3.48	3.48	3.48
13	3.06	3.21	3.30	3.35	3.38	3.41	3.42	3.44	3.45	3.45	3.46	3.46	3.47	3.47	3.47	3.47
14	3.03	3.18	3.27	3.33	3.37	3.39	3.41	3.42	3.44	3.45	3.46	3.46	3.47	3.47	3.47	3.47
15	3.01	3.16	3.25	3.31	3.36	3.38	3.40	3.42	3.43	3.44	3.45	3.46	3.47	3.47	3.47	3.47
16	3.00	3.15	3.23	3.30	3.34	3.37	3.39	3.41	3.43	3.44	3.45	3.46	3.47	3.47	3.47	3.47
17	2.98	3.13	3.22	3.28	3.33	3.36	3.38	3.40	3.42	3.44	3.45	3.46	3.47	3.47	3.47	3.47
18	2.97	3.12	3.21	3.27	3.32	3.35	3.37	3.39	3.41	3.43	3.45	3.46	3.47	3.47	3.47	3.47
19	2.96	3.11	3.19	3.26	3.31	3.35	3.37	3.39	3.41	3.43	3.44	3.46	3.47	3.47	3.47	3.47
20	2.95	3.10	3.18	3.25	3.30	3.34	3.36	3.38	3.40	3.43	3.44	3.46	3.46	3.47	3.47	3.47
22	2.93	3.08	3.17	3.24	3.29	3.32	3.35	3.37	3.39	3.42	3.44	3.45	3.46	3.47	3.47	3.47
24	2.92	3.07	3.15	3.22	3.28	3.31	3.34	3.37	3.38	3.41	3.44	3.45	3.46	3.47	3.47	3.47
26	2.91	3.06	3.14	3.21	3.27	3.30	3.34	3.36	3.38	3.41	3.43	3.45	3.46	3.47	3.47	3.47
28	2.90	3.04	3.13	3.20	3.26	3.30	3.33	3.35	3.37	3.40	3.43	3.45	3.46	3.47	3.47	3.47
30	2.89	3.04	3.12	3.20	3.25	3.29	3.32	3.35	3.37	3.40	3.43	3.44	3.46	3.47	3.47	3.47
40	2.86	3.01	3.10	3.17	3.22	3.27	3.30	3.33	3.35	3.39	3.42	3.44	3.46	3.47	3.47	3.47
60	2.83	2.98	3.08	3.14	3.20	3.24	3.28	3.31	3.33	3.37	3.40	3.43	3.45	3.47	3.48	3.48
100	2.80	2.95	3.05	3.12	3.18	3.22	3.26	3.29	3.32	3.36	3.40	3.42	3.45	3.47	3.53	3.53
∞	2.77	2.92	3.02	3.09	3.15	3.19	3.23	3.26	3.29	3.34	3.38	3.41	3.44	3.47	3.61	3.67

column labeled "3" we read down the rows until we come to the appropriate degrees of freedom. However, 57 df is not in the table. The best that we can do is to find a row for 40 df and for 60 df. The values of r_p for three groups (i.e., r_3) for 40 and 60 df are 3.01 and 2.98, respectively. By interpolating (linearly) we find that the desired value of r_3 is so close to that for 60 df that we shall use it; i.e., 2.98. This value shall be used indirectly in comparing the groups with the highest and the lowest means. More specifically, it is used for comparing the extreme means of a group of *three* means. But we will also need to compare adjacent groups. Since adjacent groups involve the comparison of only two means we need to follow the same procedure for the column labeled 2; i.e., for two groups. Reading down column 2 until we come to 40 and 60 df we find, by interpolating that r_2 is 2.83 for 57 df. These two values of r_p are written in the top row of Table 9.3. If the procedure for selecting and using the values for r_p is not clear, you should proceed through this section, accepting our statements on faith, for we discuss the matter more thoroughly later.

Table 9.3

Values of r_p and R_p for 2 and 3 groups with 57 df

Number of Groups

	2	3
r_p	2.83	2.98
R_p	2.44	2.59

Our next step is to compute the "least significant ranges" for our values, which is symbolized by R_p, where

$$(9.4) \qquad R_p = s_e r_p \sqrt{1/n}$$

Therefore, to find R_p for each number of groups we need to multiply our computed value of s_e by the appropriate value of r_p and $\sqrt{1/n}$. Since our computed value of $s_e = 3.93$, r_p for two groups is 2.83, and $n = 20$, R_2 is:

$$R_2 = (3.92)(2.83)\sqrt{1/20} = 2.44$$

Similarly R_3 is:

$$R_3 = (3.92)(2.98)\sqrt{1/20} = 2.59$$

These values of R_p have been entered in the bottom row of Table 9.3. Now let us order (rank) the means for our three groups from the lowest to the highest. They were found to be 15.05, 24.30, and 8.45 for Groups A, B, and C, respectively. Hence, our ordering is:

	Group	
C	A	B
Mean: 8.45	15.05	24.30

Now, the final step is to compare the differences between our ordered means and the values of R_p. Starting with the highest and the lowest means it can be seen that the difference between Groups B and C is 15.85. By comparing Group B with Group C we have compared the extreme means of three groups. Therefore, we read the value of $R_3 = 2.59$ from Table 9.3. This means that the difference between the group with the highest mean (Group B) and the group with the lowest mean (Group C) must exceed 2.59 for that difference to be significant at the 5 per cent level. We found that the difference between Group B and Group C was 15.85; since this value exceeds 2.59 we may conclude that Group B is significantly different from Group C.

We now should determine whether Group B is also different from Group A. Since these means are adjacent in our ranked order, we are comparing two means. Hence, we read from Table 9.3 that the appropriate value of $R_2 = 2.44$. The obtained difference between the means of these two groups is 9.25. If the obtained difference exceeds the value of 2.44, these two groups differ beyond the .05 level of significance. Since 9.25 exceeds 2.44, our conclusion is in the affirmative—Group B is significantly different from Group A.

Incidentally, we may note a rule of general procedure. If the difference between the extreme means (Groups B and C) had not been significant, there would have been no necessity for testing the lesser difference between Groups B and A. For if the extreme means did not differ significantly, any difference that is less than that (here between Groups B and A) must necessarily be *not* significant. On the other hand, if the extreme means differ significantly, the only way to find out whether the lesser differences are significant is to proceed with the test, as we have done.

The remaining comparison is between Groups A and C. We find that the difference between their means is 6.60. Since we are now comparing two groups, we read the R_p value of 2.44 from Table 9.3. As before, the necessary difference between a group of two means must exceed 2.44. Since 6.60 exceeds 2.44, we conclude that Groups A and C differ significantly.

All three groups differ significantly. Furthermore, observing the order of the means, we find that Group B has the highest mean, Group A the next, and Group C the lowest. We may then conclude that Method B led to a greater increase in reading proficiency than did Method A and that both methods A and B are superior to the normal method. Clearly, then, Method B is to be preferred.

The general rule for selecting and using the various values of r_p is: After the group means have been ordered, count the number of groups *between* the two that you are testing and add that number to those two. Then, enter the table of r_p (Table 9.2) for that number. In the three group design just discussed we have the following situation:

Group C	Group A	Group B
$\bar{X}_1$	$\bar{X}_2$	$\bar{X}_3$

If we are comparing Groups B and C, we count one group between these two, giving us a total of three. Hence, we enter column 3 of Table 9.2. A test between Groups B and A, however, has no groups intervening between them. Therefore, we add zero to our count of two, indicating that we enter column 2 of Table 9.2. And likewise, when we are comparing Groups A and C, the number is two. Let us say that we have four groups and are testing the highest mean against the lowest. In this case, two groups intervene. Therefore, we add two to the two groups that we are testing, indicating that we should enter Table 9.2 at the column labeled

4. However, we will also need to compare the highest group with the second lowest. In this case, one group will intervene, making our count three. The next comparison will be between the two highest groups, a situation in which no groups intervene. Hence, the count would be two, and we should record that value of r_p for further use. Similarly, if we have a five-groups design and are testing the highest mean against the lowest, the count would be five. For testing the highest against the second lowest, the count would be four. For testing the highest against the middle group, the count would be three. And so forth.

A Five-Groups Design

The application of Duncan's Range Test should now be clear. The same procedure is followed for experiments involving more than three groups with several minor computational differences. Its application to a design involving five groups is illustrated by the following problem. A psychologist is hired for the purpose of constructing a training program for assemblyline workers. Their job is to assemble parts of a rather complicated electronic device. Management asks the psychologist how many times the workers should practice assembling the device before they are placed on the actual production line. To be safe in answering the question the psychologist decides to conduct an experiment. He chooses to vary his independent variable, the number of times for practicing assembly of the device, in five ways: zero trials, 10 trials, 30 trials, 70 trials, and 100 trials.[6] From the fifty trainees available for his experiment, he decides to assign randomly ten to each condition. His dependent variable is the number of parts that the workers can correctly assemble in one hour, a test for which is administered after the training. Assume that the number of parts that the five groups correctly assembled are as shown in Table 9.4.

The values of ΣX, ΣX^2, n, and $\bar{X}$ have been computed for each group. To compute the sum of squares for each group we merely substitute the values required by equation 9.1:

$$SS_1 = 25 - (11)^2/10 = 12.90$$
$$SS_2 = 39 - (15)^2/10 = 16.50$$
$$SS_3 = 546 - (72)^2/10 = 27.60$$
$$SS_4 = 5533 - (235)^2/10 = 10.50$$
$$SS_5 = 5727 - (239)^2/10 = 14.90$$

[6] Actually in an experiment such as this one where the independent variable (number of trials) is quantified along some dimension, a regression analysis (such as the method of orthogonal polynomials, cf. Cochran and Cox, 1957) is more powerful. This does not mean, however, that Duncan's Range Test is inappropriate for this type of situation. Rather it means that we may fail to find differences among groups with Duncan's Range Test, when in fact they exist.

Table 9.4

Number of Parts Correctly Assembled During Test

		Group		
1	2	3	4	5

		Number of Training Trials		
0	10	30	70	100

0	2	4	24	24
1	2	5	25	24
3	1	7	23	25
0	4	9	23	25
0	0	8	25	22
2	0	7	22	24
1	0	6	24	22
0	3	8	23	26
3	2	9	22	24
1	1	9	24	23

ΣX:	11	15	72	235	239
ΣX^2:	25	39	546	5533	5727
n:	10	10	10	10	10
$\bar{X}$:	1.10	1.50	7.20	23.50	23.90

The previous formula for s_e was specialized for the case of three groups. An equation applicable to any number of groups is:

$$(9.5) \qquad s_e = \sqrt{\frac{SS_1 + SS_2 + SS_3 + \ldots SS_r}{(n_1 - 1) + (n_2 - 1) + \ldots (n_r - 1)}}$$

The numerator of this general equation for computing s_e merely indicates that you should add the sums of squares for all of the groups together. That is, you add the SS for the first group to the SS for the second group, add these to the SS for the third group, and continue in this manner until you have added the SS for all r groups, where of course r indicates the number of groups. The denominator merely indicates that you add the number of subjects in each group minus one together, continuing for all r groups. Since we have five groups in the present example, we have five sums of squares to add, and since $n_1 = n_2 = n_3 = n_4 = n_5$ we may merely multiply $n - 1$ by 5. Hence equation 9.5 may be written for this case as:

$$(9.6) \qquad s_e = \sqrt{\frac{SS_1 + SS_2 + SS_3 + SS_4 + SS_5}{5(n - 1)}}$$

Substituting the sum of squares and n for the 5 groups in equation 9.6 we obtain:

$$s_e = \sqrt{\frac{12.90 + 16.50 + 27.60 + 10.50 + 14.90}{5(10 - 1)}} = 1.35$$

The appropriate degrees of freedom is:

$$df = 50 - 5 = 45$$

For a 5 per cent level test we enter Table 9.2 to obtain the necessary values of r_p. We have five groups so we enter the column labeled 5. Reading down to 40 and 60 df we find (by interpolating) that the value of r_5 for 45 df is 3.16. Similarly the value of $r_4 = 3.10$, of $r_3 = 3.00$, and of $r_2 = 2.85$. These values are entered in the top row of Table 9.5.

Table 9.5

Values of r_p and R_p for 5 groups with 45 df

	2	3	4	5
r_p	2.85	3.00	3.10	3.16
R_p	1.23	1.30	1.34	1.37

To obtain the R_p values we multiply the appropriate value of r_p by s_e and $\sqrt{1/n}$ (equation 9.4). These computations are:

For 2 groups: $R_p = (2.85)(1.35)\sqrt{1/10} = 1.23$
" 3 " : $R_p = (3.00)(1.35)\sqrt{1/10} = 1.30$
" 4 " : $R_p = (3.10)(1.35)\sqrt{1/10} = 1.34$
" 5 " : $R_p = (3.16)(1.35)\sqrt{1/10} = 1.37$

These values have been entered in the bottom row of Table 9.5. We next order our means from lowest to highest:

		Group			
	1	2	3	4	5
Mean	1.10	1.50	7.20	23.50	23.90

Following our previous procedure we first must test the difference between the extreme means of five groups. The extreme means are for Groups 5 and 1, their difference being 22.80. Since we are considering five groups, we read the value of $R_5 = 1.37$ from Table 9.5 (for five groups: three intervening groups, plus the two we are testing). The difference of 22.80 is larger than 1.37 and, therefore, we may conclude that Group 5 is significantly different from Group 1 at the 5 per cent level. The next comparison is between Groups 5 and 2. The difference between their means is 22.40. Now we are comparing the difference between extreme values of 4 groups. Hence, we read R_p from Table 9.5 as 1.34. Since 22.40 exceeds 1.34, we may conclude that Group 5 is significantly different from Group 2. The next comparison is between Group 5 and

Group 3. Three means enter our consideration. Hence, the appropriate R_p from Table 9.5 is 1.30. The difference in means between Groups 5 and 3 is 16.70. Since this difference exceeds 1.30, these two groups are significantly different.

The difference in means between Groups 5 and 4 is 0.40. This comparison involves consideration of only two means. Therefore, the appropriate R_p is 1.23. But 0.40 does not exceed 1.23, from which we conclude that there is no significant difference between Groups 5 and 4. In short, we have found that Group 5 is significantly different from Groups 3, 2, and 1, but is *not* significantly different from Group 4. Since Group 5 has a higher mean than Groups 3, 2, and 1, we may further conclude that it is significantly superior to them. This fact may be indicated by the following scheme. Any two means that are *underscored* by the same line are *not significantly different.* Any two means that are *not underscored* by the same line *are significantly different*.[7] Since Group 5 is not significantly different from Group 4, we draw a line under those two means. But since Group 5 is significantly different from Groups 3, 2 and 1, we do not extend the line under them. That is:

	Group				
	1	2	3	4	5
Mean	1.10	1.50	7.20	<u>23.50</u>	<u>23.90</u>

We have now tested for significant differences between Group 5 and the other groups. Our next step is to compare Group 4 with Groups 3, 2, and 1. The extreme values among these four groups occur for Groups 4 and 1, their mean difference being 22.40. From Table 9.5 we find that the R_p for comparing 4 groups is 1.34. Since 22.40 is larger than the necessary value of 1.34, we may say that Groups 4 and 1 differ significantly. Now to move on to Group 2. The mean difference between Groups 4 and 2 is 22.00. From Table 9.5 we find that the appropriate value of R_p is 1.30. 22.00 exceeds 1.30. Therefore, Groups 4 and 2 differ significantly. Do Groups 4 and 3 differ significantly? The R_p for a two-groups comparison is 1.23. The difference in means of these two groups is 16.30. Therefore, the answer is in the affirmative and we may conclude that Group 4 is significantly superior to Groups 3, 2, and 1. This finding is indicated by *not* drawing a common line under Groups 4, 3, 2, and 1. We now proceed to test for significant differences between Group 3 and Groups 2 and 1. The difference between the means of Groups 3 and 1 is 6.10. From Table 9.5 we find that the value of R_p for comparing three groups is 1.30. Our obtained difference exceeds this value. We, therefore, conclude that

---

[7] It was not necessary to introduce this scheme for the preceding example since all groups were significantly different from each other. We, therefore, did not draw lines under any groups.

Group 3 is significantly different from Group 1. To compare Groups 3 and 2 we find that their mean difference is 5.70. R_p for comparing two groups is 1.23. Therefore, these two groups differ significantly. These findings are indicated by not drawing a common line under Groups 3, 2, and 1:

	Group				
	1	2	3	4	5
Mean	1.10	1.50	7.20	23.30	23.40

Our final comparison is between Groups 2 and 1. The difference between their means may be seen to be 0.40. From Table 9.5 we find a value of 1.23 for R_p when two groups are being tested. Since 0.40 does not exceed 1.23, we conclude that there is no significant difference between Groups 2 and 1. This finding is indicated by drawing a line under Groups 2 and 1:

	Group				
	1	2	3	4	5
Mean	1.10	1.50	7.20	23.20	23.40

The above lines under the means of the five groups constitute a summary of our findings. The line under Groups 4 and 5 indicates that these groups are not significantly different. But since that line does not extend to the other groups, we know that both Groups 4 and 5 are significantly superior to the other groups. The common line under Groups 1 and 2 indicates that there is no significant difference between them. The lack of a common line under Groups 3, 2, and 1 indicates that Group 3 is significantly superior to Groups 1 and 2. In short, we reach the following conclusions:

1. Groups 4 and 5 do not differ significantly.
2. Both Groups 4 and 5 are significantly superior to Groups 1, 2, and 3.
3. Group 3 is significantly superior to Groups 1 and 2.
4. Groups 1 and 2 do not differ significantly.

In Figure 9.8 we have plotted the relationship between the independent and the dependent variables. The number of practice trials used by the five groups is indicated on the horizontal axis and the mean number of parts assembled by each group in the hour test is indicated on the vertical axis. From our statistical analysis and Figure 9.8, the psychologist's answer to the problem should be obvious. Practice beyond 70 trials does not seem to increase the subject's scores, but up to that point practice is beneficial. Hence, he would recommend 70 practice trials. It may

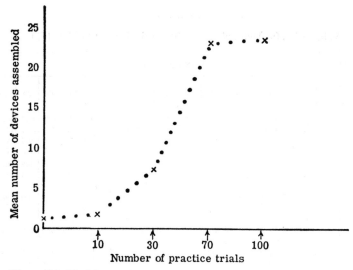

Figure 9.8. Fictitious relationship between number of practice trials and mean number of devices assembled.

be, of course, that practice beyond 100 trials would lead to a further increase in test scores, but such a possibility could only be verified by further experimentation.

In the preceding examples we have used a 5 per cent level test. Table 9.5 presents the necessary values for testing the differences between groups at the 1 per cent level. The procedure for using Table 9.5 is precisely the same as that for Table 9.2.

In order to insure that the scheme for indicating significant differences between groups is clear, let us consider several additional examples. For instance, suppose we have three groups and find that Group 3 is not significantly different from Group 2, but is significantly different from Group 1. However, we find that there is no significant difference between Groups 1 and 2. These findings would be indicated as follows:

	Group	
	1	2	3
Mean:	$\bar{X}_1$	$\bar{X}_2$	$\bar{X}_3$

Taking another example with three groups, assume that Group 3 is significantly different from Groups 2 and 1, but there is no significant difference between Groups 2 and 1:

	Group	
	1	2	3
Mean:	$\bar{X}_1$	$\bar{X}_2$	$\bar{X}_3$

Table 9.5

Values of r_p for Duncan's Range Test (Significance level $= 1\%$)

df	\multicolumn{16}{c}{Number of Groups}
	2	3	4	5	6	7	8	9	10	12	14	16	18	20	50	100
1	90.0	90.0	90.0	90.0	90.0	90.0	90.0	90.0	90.0	90.0	90.0	90.0	90.0	90.0	90.0	90.0
2	14.0	14.0	14.0	14.0	14.0	14.0	14.0	14.0	14.0	14.0	14.0	14.0	14.0	14.0	14.0	14.0
3	8.26	8.5	8.6	8.7	8.8	8.9	8.9	9.0	9.0	9.0	9.1	9.2	9.3	9.3	9.3	9.3
4	6.51	6.8	6.9	7.0	7.1	7.1	7.2	7.2	7.3	7.3	7.4	7.4	7.5	7.5	7.5	7.5
5	5.70	5.96	6.11	6.18	6.26	6.33	6.40	6.41	6.5	6.6	6.6	6.7	6.7	6.8	6.8	6.8
6	5.24	5.51	5.65	5.73	5.81	5.88	5.95	6.00	6.0	6.1	6.2	6.2	6.3	6.3	6.3	6.3
7	4.95	5.22	5.37	5.45	5.53	5.61	5.69	5.73	5.8	5.8	5.9	5.9	6.0	6.0	6.0	6.0
8	4.74	5.00	5.14	5.23	5.32	5.40	5.47	5.51	5.5	5.6	5.7	5.7	5.8	5.8	5.8	5.8
9	4.60	4.86	4.99	5.08	5.17	5.25	5.32	5.36	5.4	5.5	5.5	5.6	5.7	5.7	5.7	5.7
10	4.48	4.73	4.88	4.96	5.06	5.13	5.20	5.24	5.28	5.36	5.42	5.48	5.54	5.55	5.55	5.55
11	4.39	4.63	4.77	4.86	4.94	5.01	5.06	5.12	5.15	5.24	5.28	5.34	5.38	5.39	5.39	5.39
12	4.32	4.55	4.68	4.76	4.84	4.92	4.96	5.02	5.07	5.13	5.17	5.22	5.24	5.26	5.26	5.26
13	4.26	4.48	4.62	4.69	4.74	4.84	4.88	4.94	4.98	5.04	5.08	5.13	5.14	5.15	5.15	5.15
14	4.21	4.42	4.55	4.63	4.70	4.78	4.83	4.87	4.91	4.96	5.00	5.04	5.06	5.07	5.07	5.07
15	4.17	4.37	4.50	4.58	4.64	4.72	4.77	4.81	4.84	4.90	4.94	4.97	4.99	5.00	5.00	5.00
16	4.13	4.34	4.45	4.54	4.60	4.67	4.72	4.76	4.79	4.84	4.88	4.91	4.93	4.94	4.94	4.94
17	4.10	4.30	4.41	4.50	4.56	4.63	4.68	4.72	4.75	4.80	4.83	4.86	4.88	4.89	4.89	4.89
18	4.07	4.27	4.38	4.46	4.53	4.59	4.64	4.68	4.71	4.76	4.79	4.82	4.84	4.85	4.85	4.85
19	4.05	4.24	4.35	4.43	4.50	4.56	4.61	4.64	4.67	4.72	4.76	4.79	4.81	4.82	4.82	4.82
20	4.02	4.22	4.33	4.40	4.47	4.53	4.58	4.61	4.65	4.69	4.73	4.76	4.78	4.79	4.79	4.79
22	3.99	4.17	4.28	4.36	4.42	4.48	4.53	4.57	4.60	4.65	4.68	4.71	4.74	4.75	4.75	4.75
24	3.96	4.14	4.24	4.33	4.39	4.44	4.49	4.53	4.57	4.62	4.64	4.67	4.70	4.72	4.74	4.74
26	3.93	4.11	4.21	4.30	4.36	4.41	4.46	4.50	4.53	4.58	4.62	4.65	4.67	4.69	4.73	4.73
28	3.91	4.08	4.18	4.28	4.34	4.39	4.43	4.47	4.51	4.56	4.60	4.62	4.65	4.67	4.72	4.72
30	3.89	4.06	4.16	4.22	4.32	4.36	4.41	4.45	4.48	4.54	4.58	4.61	4.63	4.65	4.71	4.71
40	3.82	3.99	4.10	4.17	4.24	4.30	4.34	4.37	4.41	4.46	4.51	4.54	4.57	4.59	4.69	4.69
60	3.76	3.92	4.03	4.12	4.17	4.23	4.27	4.31	4.34	4.39	4.44	4.47	4.50	4.53	4.66	4.66
100	3.71	3.86	3.98	4.06	4.11	4.17	4.21	4.25	4.29	4.35	4.38	4.42	4.45	4.48	4.64	4.65
∞	3.64	3.80	3.90	3.98	4.04	4.09	4.14	4.17	4.20	4.26	4.31	4.34	4.38	4.41	4.60	4.63

Consider a four-groups design. Assume that Group 4 is significantly superior to the other groups; that Group 3 is not significantly superior to Group 2, but is significantly superior to Group 1; and that there is no difference between Groups 1 and 2:

		Group		
	1	2	3	4
Mean:	$\overline{X}_1$	$\overline{X}_2$	$\overline{X}_3$	$\overline{X}_4$

With four groups again, assume that Group 4 does not differ significantly from Groups 3 and 2, but is significantly different from Group 1. Furthermore, Groups 3 and 2 are not significantly different, but Group 3 is significantly superior to Group 1. And Group 2 is significantly different from Group 1:

Group

	1	2	3	4
Mean:	$\bar{X}_1$	$\bar{X}_2$	$\bar{X}_3$	$\bar{X}_4$

In the above example the top line indicates that Groups 2, 3, and 4 do not differ significantly. The bottom line indicates that Groups 2 and 3 do not differ significantly. Since the latter information is contained in the former, the bottom line need not be drawn. Therefore, this example should be:

Group

	1	2	3	4
Mean:	$\bar{X}_1$	$\bar{X}_2$	$\bar{X}_3$	$\bar{X}_4$

Statistical Analysis for Unequal n's

The preceding discussion has assumed that the number of subjects in each of the several groups is equal. However, it is frequently the case that different numbers of subjects are assigned to the groups. Kramer (1956) has extended Duncan's Range Test so that it is applicable to this situation [see also Duncan (1957) and Kramer (1957) for this and additional extensions]. The same general procedure as that for equal ns is followed with only minor exceptions. To illustrate, assume that we wish to test the differences between means for the set of data in Table 9.6.

Table 9.6

Scores Obtained in a Three Group Experiment with Unequal ns

Group 1	Group 2	Group 3
3	4	10
4	3	12
5	5	12
4	2	14
3	1	15
1	3	17
7	7	13
6	9	11
$\Sigma X = 33$	1	12
$\Sigma X^2 = 161$	1	11
$n = 8$	2	15
$\bar{X} = 4.12$	$\Sigma X = 38$	14
	$\Sigma X^2 = 200$	12
	$n = 11$	13
	$\bar{X} = 3.45$	$\Sigma X = 181$
		$\Sigma X^2 = 2387$
		$n = 14$
		$\bar{X} = 12.93$

The first step is to compute ΣX, ΣX^2, n and $\overline{X}$ for each group (indicated above). With these values we shall compute the sum of squares of the three groups:

$$SS_1 = 161 - (33)^2/8 = 24.88$$
$$SS_2 = 200 - (38)^2/11 = 68.73$$
$$SS_3 = 2387 - (181)^2/14 = 46.93$$

We need next to compute s_e. However since the ns of the groups are not equal, we shall have to use equation 9.5 which, for three groups, becomes:

$$(9.7) \qquad s_e = \sqrt{\frac{SS_1 + SS_2 + SS_3}{(n_1 - 1) + (n_2 - 1) + (n_3 - 1)}}$$

Substituting the appropriate values of SS and n in equation 9.7:

$$s_e = \sqrt{\frac{24.88 + 68.73 + 46.93}{(8 - 1) + (11 - 1) + (14 - 1)}} = 2.16$$

The df are computed as before:

$$df = 33 - 3 = 30$$

We now write down the values of r_p for three and two groups. Assuming a 5 per cent level test we enter the column marked 3 in Table 9.2. Reading down the column we find the r_p for three groups that corresponds to 30 df. It is 3.04. For two groups $r_p = 2.89$.

Table 9.7

Values of r_p for a Three-Groups Design with 30 df

	Groups	
	2	3
r_p	2.89	3.04

Up to this point the procedure for unequal ns has been essentially the same as for equal ns. We shall now depart to some extent. Instead of using equation 9.4 we now use equation 9.8 for computing R_p when we have unequal ns:

$$(9.8) \qquad R_p = (s_e)(r_p)\sqrt{\tfrac{1}{2}(1/n_a + 1/n_b)}$$

Where n_a and n_b are the ns for whatever two groups are being compared, as we shall shortly see. Let us start by ordering the means and testing the difference between the extreme groups:

	Group		
	2	1	3
Mean:	3.45	4.12	12.93

The difference in means between Group 3 and Group 2 is 9.48. We next determine the value of R_p for this test. s_e will be the same for all tests, but r_p will depend on the number of groups being compared. In this comparison we are considering three means, so r_p will be that for comparing the extreme means of three groups. From Table 9.7 we read: $r_3 = 3.04$. The ns are those for groups 2 and 3, i.e., 11 and 14 respectively. Hence, R_p is:

$$R_p = (2.16)(3.04)\sqrt{\frac{1}{2}\left(\frac{1}{11} + \frac{1}{14}\right)} = (6.57)\sqrt{.0812}$$

$$= (6.57)(.28) = 1.84$$

Since the mean difference between Groups 3 and 2 (9.48) exceeds the R_p for this comparison (1.84), we may conclude that Group 3 is significantly superior to Group 2. Now to test between the means of Groups 3 and 1. r_p for two groups, read from Table 9.7, is 2.89. The ns for Groups 3 and 1 are 14 and 8, respectively. Using the same value of s_e and substituting these values in equation 9.8, R_p for this comparison is calculated as follows:

$$R_p = (2.16)(2.89)\sqrt{\tfrac{1}{2}(1/14 + 1/8)} = 1.93$$

The difference between the means of Groups 3 and 1 is 8.81. Since this value exceeds the necessary value of R_p (1.93), Group 3 is significantly superior to Group 1.

The final comparison is between Groups 1 and 2. The difference between their means is 0.67. We are comparing two adjacent groups. Hence, r_2 may be read from Table 9.7 as 2.89. The ns for these two groups are 8 and 11. With the same s_e as before, the R_p for this test is:

$$R_p = (2.16)(2.89)\sqrt{\tfrac{1}{2}(1/8 + 1/11)} = 2.06$$

Since the mean difference between Groups 1 and 2 (.67) does not exceed the necessary value of R_p (2.06), we may conclude that there is no significant difference between these two groups.

The statistical findings may be summarized as follows:

	Group		
	2	1	3
Mean:	3.45	4.12	12.93

The lack of a line under Group 3 indicates that it is significantly superior to Groups 1 and 2. The common underline beneath Groups 1 and 2 indicates that there is no significant difference between them.

The above procedure can be extended to any number of groups. Besides substituting the appropriate ns in equation 9.8, one merely needs to be cautious about selecting the appropriate value of r_p.

Analysis of Variance

Now having discussed Duncan's Range Test, we shall consider the second (and more commonly used) type of statistical procedure applied to the multi-groups design considered in this chapter. However, we shall attempt to explain only enough features of this procedure to allow you to apply it to experimental work.[8]

As with the preceding designs, we set up a null hypothesis. Our statistical test will allow us to either reject the null hypothesis or fail to reject it. For the present design, consider the null hypothesis to be that there is no difference among the means of the several groups. Keeping this null hypothesis in mind, let us return to it after our consideration of analysis of variance.

We already have some acquaintance with the term *variance* (p. 154) which will help us in the ensuing discussion. Review it now.

The simplest application of analysis of variance would be in testing the difference between two groups. We have already discussed the *t* test for this purpose. However, equivalent results would be obtained by applying analysis of variance to a two-group design. That is, we could analyze a two-groups design by using either the *t* test or the technique of analysis of variance and obtain precisely the same conclusions. The same statement cannot be made if more than two groups are used, for in this case the *t* test is simply not appropriate. Let us say that the dependent variable scores that result from a two-groups design are those plotted in Figure 9.9. That is, the curve to the left represents the scores made by the subjects in Group 1 while the frequency distribution to the right is for Group 2.

Now, are these groups significantly different? To answer this question by using analysis of variance we first need to note that *total variance* may be determined.[9] The total variance is the variance that results when we take all subjects in the experiment into account as one sample: the total variance is the variance computed from the dependent variable scores of *all* the subjects, ignoring the fact that some were under one experimental condition while others were under another experimental condition. The important point for us to observe here is that the total variance may be partitioned (analyzed) into parts (hence the term, "analysis of variance"). In particular, the total variance may be partitioned into two major components: the variance *between groups* and

[8] For a simplified but more thorough treatment of analysis of variance by psychologists you should read Underwood et al. (1954), or for more detailed treatments see Edwards (1950) or McNemar (1955).
[9] We are approaching analysis of variance by talking about population values. While this is an exception to our general approach, it is justified on pedagogical grounds. We shall consider sample values shortly when we talk about sums of squares.

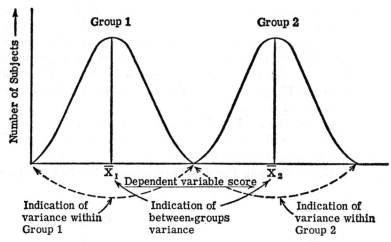

Figure 9.9. A crude indication of the nature of within- and between-groups variances using only two groups.

that *within groups*. *Roughly*, the variance between groups may be thought of as determined by the extent to which the means of the two groups differ.

In Figure 9.9 the between-groups variance is crudely indicated by the difference between the two means. More accurately we may say that the larger the difference between the means the larger is the between-groups variance. The within-groups variance, on the other hand, is determined by the extent to which the subjects in each group differ. If the subjects in Group 1 differ sizeably among themselves, and if the same is true for subjects in Group 2, the within-groups variance is going to be large. And the larger the within-groups variance, the larger the "error" variance in the experiment. By way of illustration, assume that all of the subjects in Group 1 have been treated precisely alike. Hence, if they were precisely alike when they went into the experiment they should all receive the same score on the dependent variable. If this happened, the within-groups variance would be zero—for there would be no variation among their scores. Of course, the within-groups variance is almost never zero, since all the subjects are not the same before the experiment and the experimenter is never able to treat all of them precisely alike.

Let us now reason by analogy to the *t* test. You will recall that the numerator of equation 5.2 measures the difference between groups. It is thus analogous to our between-groups variance. The denominator of equation 5.2 is a measure of the "error" in the experiment, and is thus analogous to our within-groups variance. This should be apparent when one notes that the denominator of equation 5.2 is large if the variances of the groups are large, and small if the variances of the groups are small (see p. 289). Recall that the larger the numerator and the smaller the denominator of the t ratio, the greater the likelihood that the two groups are significantly different. The same is true in our analogy: the larger the between-groups variance and the smaller the within-groups variance, the more likely our groups are to be significantly different. Looking at Figure 9.9 we may say that the larger the distance between the two means, and the smaller the within (internal) variances of the two groups, the more likely they are to be significantly different. For example, the difference between the two groups of Figure 9.10 is more likely to be significant than the difference between the two groups of

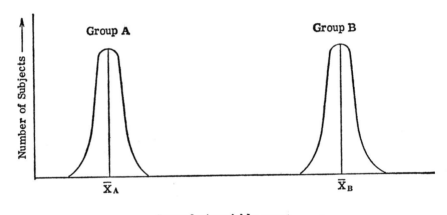

Dependent variable score ──►

Figure 9.10. A more extreme difference between two groups than that shown in Figure 9.9. Here the between-groups variance is greater but the within-groups variance is less.

Figure 9.9. This is so because the difference between the means in Figure 9.10 is represented as greater than that for Figure 9.9, and also because the variance within the groups of Figure 9.10 is less than for Figure 9.9.

We have discussed the case of two groups. Precisely the same general reasoning applies when there are more than two groups: the total variance in the experiment is analyzed into two parts, the within- and the between-groups variances. If the difference among the several means is large, the between-groups variance will be large. If the difference among the several means is small, the between-groups variance will be small. If the subjects who are treated alike differ sizeably, then the within (internal) variance of each group will be large. And if the individual group variances are large, the within-group variances will be large. The larger the between-groups variance and the smaller the within-group variance, the more likely it is that the groups differ significantly.

We have said that the total variance can be analyzed into two components. We are not immediately concerned with variances, however, but with certain sums of squares (SS), which are used to compute variances. We shall compute the total SS and then analyze it into its parts. A generalized equation for computing the total SS is:

$$(9.9) \quad \text{Total } SS = (\Sigma X_1^2 + \Sigma X_2^2 + \Sigma X_3^2 \ldots + \Sigma X_r^2)$$
$$- \frac{(\Sigma X_1 + \Sigma X_2 + \Sigma X_3 + \ldots + \Sigma X_r)^2}{N}$$

where, as before, the subscript r simply indicates that we continue adding the values indicated (the sum of X-squares, and the sum of X respectively) for as many groups as we have in the experiment.

Our next step is to analyze the total SS into components. There are two major components, that between groups and that within groups. A generalized equation for computing the between-groups SS is:

$$(9.10) \quad \text{Between } SS = \frac{(\Sigma X_1)^2}{n_1} + \frac{(\Sigma X_2)^2}{n_2} + \frac{(\Sigma X_3)^2}{n_3} \ldots \frac{(\Sigma X_r)^2}{n_r}$$
$$- \frac{(\Sigma X_1 + \Sigma X_2 + \Sigma X_3 \ldots \Sigma X_r)^2}{N}$$

The within-groups component of the total SS may be computed by subtraction. That is,

$$(9.11) \quad \text{Within } SS = \text{Total } SS - \text{Between } SS$$

In a more-than-two-randomized-groups design, of course, there may be any number of groups. To compute the several SS we must compute the ΣX and ΣX^2 separately for each group. The subscripts, as before, indicate the different groups. Hence ΣX_1 is the sum of the dependent variable scores for Group 1, ΣX_3^2 is the sum of the squares of the dependent variable scores for Group 3, and so forth. And N remains the total number of subjects in the experiment. To illustrate the analysis of variance procedure, consider the data in Table 9.8.

To compute the total SS, we may write the specialized form of equation 9.9 for three groups as follows:

$$(9.12) \quad \text{Total } SS = (\Sigma X_1^2 + \Sigma X_2^2 + \Sigma X_3^2) - \frac{(\Sigma X_1 + \Sigma X_2 + \Sigma X_3)^2}{N}$$

We can see that the sum of X for Group 1 is 9, for Group 2 it is 20, and for Group 3 it is 30. Or written in terms of equation 9.12 we may say that $\Sigma X_1 = 9$, $\Sigma X_2 = 20$, and $\Sigma X_3 = 30$. Similarly, $\Sigma X_1^2 = 35$, $\Sigma X_2^2 = 106$, $\Sigma X_3^2 = 236$, and $N = 12$. Substituting these values in equation 9.12, we find the total SS to be:

Table 9.8

Fictitious Dependent Variable Scores
for a Three-Randomized-Groups Design

	Group	
1	2	3
0	3	5
1	5	7
3	6	9
5	6	9
ΣX: 9	20	30
ΣX^2: 35	106	236
n: 4	4	4

$$\text{Total } SS = (35 + 106 + 236) - \frac{(9 + 20 + 30)^2}{12} = 86.92$$

To compute the between-groups SS for three groups, we substitute the appropriate values in equation 9.13, the specialized form of equation 9.10 for three groups. This requires that we merely substitute the value of ΣX for each group, square it, and divide by the number of subjects in each group. The last term, we may note, is the same as the last term in equation 9.12. For this reason it is not necessary to compute it again, providing there was no error in its computation the first time. Making the appropriate substitutions from Table 9.8, and performing the indicated computations we find that:

(9.13)

$$\text{Between groups } SS = \frac{(\Sigma X_1)^2}{n_1} + \frac{(\Sigma X_2)^2}{n_2} + \frac{(\Sigma X_3)^2}{n_3} - \frac{(\Sigma X_1 + \Sigma X_2 + \Sigma X_3)^2}{N}$$

$$= (9)^2/4 + (20)^2/4 + (30)^2/4 - 290.08 = 55.17$$

Substituting in equation 9.11, we find that:

$$\text{Within-groups } SS = 86.92 - 55.17 = 31.75$$

We have previously mentioned variances a number of times. You may have wondered where the variances are that we analyze when we conduct an analysis of variance. We shall consider them now, but under a different name, for they are referred to in analysis of variance procedures not as variances, but as *mean squares*. The rule for computing mean squares is simple: divide a given sum of squares by the appropriate degrees of freedom.

In introducing the equations that we use to determine the three degrees of freedom that we need, let us emphasize what we have done with regard to sums of squares—we have computed a total SS, and partitioned it into two parts, the between SS and the within SS. The same procedure is followed for df. First we determine that:

(9.14) $$\text{Total } df = N - 1$$

then that:

(9.15) $$\text{Between } df = r - 1$$

and that:

(9.16) $$\text{Within } df = N - r$$

For our example we then find that, with $N = 12$, and $r = 3$,

$$\text{Total } df = 12 - 1 = 11$$
$$\text{Between } df = 3 - 1 = 2$$
$$\text{Within } df = 12 - 3 = 9$$

And we may note that the between df plus the within df add up to equal the total df $(9 + 2 = 11)$.

There are two mean squares that we need to compute, a mean square for our between groups source of variation, and that for our within groups. To compute the former we divide the between groups SS by the between groups df, and similarly for the latter. Hence, the within groups mean square is 31.75 divided by 9. We shall enter these values in a summary table (Table 9.9).

Table 9.9

Summary Table for an Analysis of Variance

Source of Variation	Sum of Squares	df	Mean Square	F
Between Groups	55.17	2	27.58	7.81
Within Groups	31.75	9	3.53	
Total	86.93	11		

Now, as we have previously indicated, if our between groups mean square is sizeable, relative to our within groups mean square, then we may conclude that the dependent variable values for our groups are different.

However, we must again face the problem: how sizeable is "sizeable," i.e., how large must the between component be in order for us to conclude that a given independent variable is effective? To answer this we apply a suitable statistical test. The test that is considered most appropriate is the F test, which was developed by Professor Sir Ronald Aymer Fisher, one of the outstanding statisticians of all time. It was so named in his honor by another outstanding statistician, Professor George W. Snedecor. The F test is defined as follows:

$$(9.17) \qquad F = \frac{\text{Mean square between groups}}{\text{Mean square within groups}}$$

This test is obviously easy to conduct. And we may note the similarity between it and the t test. For in both cases the numerator is an indication of the differences between groups and the denominator is an indication of the experimental error, or as it is also called, the error variance of the experiment. More particularly, in the simplest applications of the F test, the numerator contains an estimate of the error variance plus an estimate of the "real" effect (if any) of the independent variable. The denominator is only an *estimate* of the error variance. Now you can see what happens when you divide the numerator by the denominator: the computed value of F reflects the effect of the independent variable. For example, suppose the independent variable is totally ineffective in influencing the dependent variable. In this case we would expect (at least in the long-run) that the numerator would not contain any contribution from the independent variable (there would be no "real" between-groups variance). Hence, the value for the numerator would only be our estimate of the error variance. And a similar estimate of the error variance is in the denominator. Therefore, in the long-run if you divide the value for the error variance by the value for the error variance, you obtain an F of 1.0.

This reasoning is made apparent by rewriting the equation for F as follows:

$$(9.18) \qquad F = \frac{\left(\begin{array}{c}\text{Estimate of the}\\ \text{error variance}\end{array}\right) + \left(\begin{array}{c}n \text{ times the estimate of the}\\ \text{effect of the independent variable}\end{array}\right)}{\text{Estimate of the error variance}}$$

Now suppose that the estimates of the error variance in the numerator and in the denominator are the same. Suppose they are 5. If the estimate of the effect of the independent variable is 3, in this case for $n = 5$, F would be:

$$F = \frac{5 + (5)(3)}{5} = 4.00$$

Thus, since the apparent effect of n times the independent variable plus the error estimate is considerably larger than the error estimate in the denominator, we would suspect that the independent variable was effective. On the other hand, if the estimated effect of our independent variable is zero, F would be:

$$F = \frac{5 + 0}{5} = 1.0$$

Thus, any time we obtain an F of 1.0 we are rather certain that chance and chance alone has influenced our dependent variable. However, we might obtain some apparent effect of our independent variable by chance, just as two means might differ to some extent by chance. If this happens, then our numerator will be somewhat larger than our denominator—the between-groups mean square will be somewhat larger than we should expect by chance. Now the question is, how large can the numerator of the F ratio be before it is too large to be the result of chance variation? For, if the numerator is large, the value of F will be large. You will note that this is the same question we asked concerning the t test. That is, how large must t be before we can reject chance as an explanation of our results? And we shall answer this question for the F test in a manner similar to that for the t test. First, however, we should actually compute our values of F. Following our example, we have divided the mean square within groups (3.53) into the mean square between groups (27.58), and inserted the resulting value of F in Table 9.9.

Just as with the t test, we next must determine the value of P that is associated with our computed value of F. Assuming that we have set a significance level of 0.05, if our value of F has a probability of less than 0.05, then we may reject our null hypothesis—we may assert that there is a significant difference among our groups. If, however, the P associated with our F is larger than 0.05, then we fail to reject the null hypothesis—we conclude that there is no difference among our groups.

To ascertain the value of P associated with our F, refer to Table 9.10 which fills the same function as the Table of t, although it is a bit different to use. Let us initially note that (1) across the top we find "df associated with the numerator" and (2) down the left side we find "df associated with the denominator." Therefore, we know that we need two df to enter the table of F. In this example we have two df for between groups (the numerator of the F test) and 9 df for within groups (the denominator of the F test). Hence we find the column labeled "2" and read down to the row labeled "9". We find a row for a P of 0.01, a row for a P of 0.05, and rows for Ps of 0.10 and 0.20. We are making a 0.05 level test, so we shall ignore the other values of P. With two and nine df, we find that we must have an F of 4.26 for significance at the 5 per cent level. Since our computed F(7.81) exceeds this value, we reject our null hypothesis—we conclude that our groups differ significantly. And on the assumption that proper experimental techniques have obtained in our hypothetical experiment, we conclude that variation of our independent variable significantly influenced our dependent variable.

Just as with the Table of t, you should study the Table of F sufficiently to make sure that you have an adequate understanding of it. To provide a little practice, say that you have six groups in your experiment with 10 subjects per group. In this event you have five degrees of freedom for your between source of variation and 54 df for the

Table 9.10*

Table of F

df Associated with Denominator	P	\multicolumn{9}{c	}{df Associated with Numerator}
		1	2	3	4	5	6	8	12	24	∞
1	0.01	4052	4999	5403	5625	5764	5859	5981	6106	6234	6366
	0.05	161.45	199.50	215.71	224.58	230.16	233.99	238.88	243.91	249.05	254.32
	0.10	39.86	49.50	53.59	55.83	57.24	58.20	59.44	60.70	62.00	63.33
	0.20	9.47	12.00	13.06	13.73	14.01	14.26	14.59	14.90	15.24	15.58
2	0.01	98.49	99.00	99.17	99.25	99.30	99.33	99.36	99.42	99.46	99.50
	0.05	18.51	19.00	19.16	19.25	19.30	19.33	19.37	19.41	19.45	19.50
	0.10	8.53	9.00	9.16	9.24	9.29	9.33	9.37	9.41	9.45	9.49
	0.20	3.56	4.00	4.16	4.24	4.28	4.32	4.36	4.40	4.44	4.48
3	0.01	34.12	30.81	29.46	28.71	28.24	27.91	27.49	27.05	26.60	26.12
	0.05	10.13	9.55	9.28	9.12	9.01	8.94	8.84	8.74	8.64	8.53
	0.10	5.54	5.46	5.39	5.34	5.31	5.28	5.25	5.22	5.18	5.13
	0.20	2.68	2.89	2.94	2.96	2.97	2.97	2.98	2.98	2.98	2.98
4	0.01	21.20	18.00	16.69	15.98	15.52	15.21	14.80	14.37	13.93	13.46
	0.05	7.71	6.94	6.59	6.39	6.26	6.16	6.04	5.91	5.77	5.63
	0.10	4.54	4.32	4.19	4.11	4.05	4.01	3.95	3.90	3.83	3.76
	0.20	2.35	2.47	2.48	2.48	2.48	2.47	2.47	2.46	2.44	2.43
5	0.01	16.26	13.27	12.06	11.39	10.97	10.67	10.29	9.89	9.47	9.02
	0.05	6.61	5.79	5.41	5.19	5.05	4.95	4.82	4.68	4.53	4.36
	0.10	4.06	3.78	3.62	3.52	3.45	3.40	3.34	3.27	3.19	3.10
	0.20	2.18	2.26	2.25	2.24	2.23	2.22	2.20	2.18	2.16	2.13
6	0.01	13.74	10.92	9.78	9.15	8.75	8.47	8.10	7.72	7.31	6.88
	0.05	5.99	5.14	4.76	4.53	4.39	4.28	4.15	4.00	3.84	3.67
	0.10	3.78	3.46	3.29	3.18	3.11	3.05	2.98	2.90	2.82	2.72
	0.20	2.07	2.13	2.11	2.09	2.08	2.06	2.04	2.02	1.99	1.95
7	0.01	12.25	9.55	8.45	7.85	7.46	7.19	6.84	6.47	6.07	5.65
	0.05	5.59	4.74	4.35	4.12	3.97	3.87	3.73	3.57	3.41	3.23
	0.10	3.59	3.26	3.07	2.96	2.88	2.83	2.75	2.67	2.58	2.47
	0.20	2.00	2.04	2.02	1.99	1.97	1.96	1.93	1.91	1.87	1.83
8	0.01	11.26	8.65	7.59	7.01	6.63	6.37	6.03	5.67	5.28	4.86
	0.05	5.32	4.46	4.07	3.84	3.69	3.58	3.44	3.28	3.12	2.93
	0.10	3.46	3.11	2.92	2.81	2.73	2.67	2.59	2.50	2.40	2.29
	0.20	1.95	1.98	1.95	1.92	1.90	1.88	1.86	1.83	1.79	1.74
9	0.01	10.56	8.02	6.99	6.42	6.06	5.80	5.47	5.11	4.73	4.31
	0.05	5.12	4.26	3.86	3.63	3.48	3.37	3.23	3.07	2.90	2.71
	0.10	3.36	3.01	2.81	2.69	2.61	2.55	2.47	2.38	2.28	2.16
	0.20	1.91	1.94	1.90	1.87	1.85	1.83	1.80	1.76	1.72	1.67
10	0.01	10.04	7.56	6.55	5.99	5.64	5.39	5.06	4.71	4.33	3.91
	0.05	4.96	4.10	3.71	3.48	3.33	3.22	3.07	2.91	2.74	2.54
	0.10	3.28	2.92	2.73	2.61	2.52	2.46	2.38	2.28	2.18	2.06
	0.20	1.88	1.90	1.86	1.83	1.80	1.78	1.75	1.72	1.67	1.62
11	0.01	9.65	7.20	6.22	5.67	5.32	5.07	4.74	4.40	4.02	3.60
	0.05	4.84	3.98	3.59	3.36	3.20	3.09	2.95	2.79	2.61	2.40
	0.10	3.23	2.86	2.66	2.54	2.45	2.39	2.30	2.21	2.10	1.97
	0.20	1.86	1.87	1.83	1.80	1.77	1.75	1.72	1.68	1.63	1.57

Table 9.10* (*Continued*)

df Associated with Denominator		*df* Associated with Numerator									
	P	1	2	3	4	5	6	8	12	24	∞
12											
	0.01	9.33	6.93	5.95	5.41	5.06	4.82	4.50	4.16	3.78	3.36
	0.05	4.75	3.88	3.49	3.26	3.11	3.00	2.85	2.69	2.50	2.30
	0.10	3.18	2.81	2.61	2.48	2.39	2.33	2.24	2.15	2.04	1.90
	0.20	1.84	1.85	1.80	1.77	1.74	1.72	1.69	1.65	1.60	1.54
13											
	0.01	9.07	6.70	5.74	5.20	4.86	4.62	4.30	3.96	3.59	3.16
	0.05	4.67	3.80	3.41	3.18	3.02	2.92	2.77	2.60	2.42	2.21
	0.10	3.14	2.76	2.56	2.43	2.35	2.28	2.20	2.10	1.98	1.85
	0.20	1.82	1.83	1.78	1.75	1.72	1.69	1.66	1.62	1.57	1.51
14											
	0.01	8.86	6.51	5.56	5.03	4.69	4.46	4.14	3.80	3.43	3.00
	0.05	4.60	3.74	3.34	3.11	2.96	2.85	2.70	2.53	2.35	2.13
	0.10	3.10	2.73	2.52	2.39	2.31	2.24	2.15	2.05	1.94	1.80
	0.20	1.81	1.81	1.76	1.73	1.70	1.67	1.64	1.60	1.55	1.48
15											
	0.01	8.68	6.36	5.42	4.89	4.56	4.32	4.00	3.67	3.29	2.87
	0.05	4.54	3.68	3.29	3.06	2.90	2.79	2.64	2.48	2.29	2.07
	0.10	3.07	2.70	2.49	2.36	2.27	2.21	2.12	2.02	1.90	1.76
	0.20	1.80	1.79	1.75	1.71	1.68	1.66	1.62	1.58	1.53	1.46
16											
	0.01	8.53	6.23	5.29	4.77	4.44	4.20	3.89	3.55	3.18	2.75
	0.05	4.49	3.63	3.24	3.01	2.85	2.74	2.59	2.42	2.24	2.01
	0.10	3.05	2.67	2.46	2.33	2.24	2.18	2.09	1.99	1.87	1.72
	0.20	1.79	1.78	1.74	1.70	1.67	1.64	1.61	1.56	1.51	1.43
17											
	0.01	8.40	6.11	5.18	4.67	4.34	4.10	3.79	3.45	3.08	2.65
	0.05	4.45	3.59	3.20	2.96	2.81	2.70	2.55	2.38	2.19	1.96
	0.10	3.03	2.64	2.44	2.31	2.22	2.15	2.06	1.96	1.84	1.69
	0.20	1.78	1.77	1.72	1.68	1.65	1.63	1.59	1.55	1.49	1.42
18											
	0.01	8.28	6.01	5.09	4.58	4.25	4.01	3.71	3.37	3.00	2.57
	0.05	4.41	3.55	3.16	2.93	2.77	2.66	2.51	2.34	2.15	1.92
	0.10	3.01	3.62	2.42	2.29	2.20	2.13	2.04	1.93	1.81	1.66
	0.20	1.77	1.76	1.71	1.67	1.64	1.62	1.58	1.53	1.48	1.40
19											
	0.01	8.18	5.93	5.01	4.50	4.17	3.94	3.63	3.30	2.92	2.49
	0.05	4.38	3.52	3.13	2.90	2.74	2.63	2.48	2.31	2.11	1.88
	0.10	2.99	2.61	2.40	2.27	2.18	2.11	2.02	1.91	1.79	1.63
	0.20	1.76	1.75	1.70	1.66	1.63	1.61	1.57	1.52	1.46	1.39
20											
	0.01	8.10	5.85	4.94	4.43	4.10	3.87	3.56	3.23	2.86	2.42
	0.05	4.35	3.49	3.10	2.87	2.71	2.60	2.45	2.28	2.08	1.84
	0.10	2.97	2.59	2.38	2.25	2.16	2.09	2.00	1.89	1.77	1.61
	0.20	1.76	1.75	1.70	1.65	1.62	1.60	1.56	1.51	1.45	1.37
21											
	0.01	8.02	5.78	4.87	4.37	4.04	3.81	3.51	3.17	2.80	2.36
	0.05	4.32	3.47	3.07	2.84	2.68	2.57	2.42	2.25	2.05	1.81
	0.10	2.96	2.57	2.36	2.23	2.14	2.08	1.98	1.88	1.75	1.59
	0.20	1.75	1.74	1.69	1.65	1.61	1.59	1.55	1.50	1.44	1.36
22											
	0.01	7.94	5.72	4.82	4.31	3.99	3.76	3.45	3.12	2.75	2.31
	0.05	4.30	3.44	3.05	2.82	2.66	2.55	2.40	2.23	2.03	1.78
	0.10	2.95	2.56	2.35	2.22	2.13	2.06	1.97	1.86	1.73	1.57
	0.20	1.75	1.73	1.68	1.64	1.61	1.58	1.54	1.49	1.43	1.35

Table 9.10* (Continued)

df Associated with Denominator	P	1	2	3	4	5	6	8	12	24	∞
23											
	0.01	7.88	5.66	4.76	4.26	3.94	3.71	3.41	3.07	2.70	2.26
	0.05	4.28	3.42	3.03	2.80	2.64	2.53	2.38	2.20	2.00	1.76
	0.10	2.94	2.55	2.34	2.21	2.11	2.05	1.95	1.84	1.72	1.55
	0.20	1.74	1.73	1.68	1.63	1.60	1.57	1.53	1.49	1.42	1.34
24											
	0.01	7.82	5.61	4.72	4.22	3.90	3.67	3.36	3.03	2.66	2.21
	0.05	4.26	3.40	3.01	2.78	2.62	2.51	2.36	2.18	1.98	1.73
	0.10	2.93	2.54	2.33	2.19	2.10	2.04	1.94	1.83	1.70	1.53
	0.20	1.74	1.72	1.67	1.63	1.59	1.57	1.53	1.48	1.42	1.33
25											
	0.01	7.77	5.57	4.68	4.18	3.86	3.63	3.32	2.99	2.62	2.17
	0.05	4.24	3.38	2.99	2.76	2.60	2.49	2.34	2.16	1.96	1.71
	0.10	2.92	2.53	2.32	2.18	2.09	2.02	1.93	1.82	1.69	1.52
	0.20	1.73	1.72	1.66	1.62	1.59	1.56	1.52	1.47	1.41	1.32
26											
	0.01	7.72	5.53	4.64	4.14	3.82	3.59	3.29	2.96	2.58	2.13
	0.05	4.22	3.37	2.98	2.74	2.59	2.47	2.32	2.15	1.95	1.69
	0.10	2.91	2.52	2.31	2.17	2.08	2.01	1.92	1.81	1.68	1.50
	0.20	1.73	1.71	1.66	1.62	1.58	1.56	1.52	1.47	1.40	1.31
27											
	0.01	7.68	5.49	4.60	4.11	3.78	3.56	3.26	2.93	2.55	2.10
	0.05	4.21	3.35	2.96	2.73	2.57	2.46	2.30	2.13	1.93	1.67
	0.10	2.90	2.51	2.30	2.17	2.07	2.00	1.91	1.80	1.67	1.49
	0.20	1.73	1.71	1.66	1.61	1.58	1.55	1.51	1.46	1.40	1.30
28											
	0.01	7.64	5.45	4.57	4.07	3.75	3.53	3.23	2.90	2.52	2.06
	0.05	4.20	3.34	2.95	2.71	2.56	2.44	2.29	2.12	1.91	1.65
	0.10	2.89	2.50	2.29	2.16	2.06	2.00	1.90	1.79	1.66	1.48
	0.20	1.72	1.71	1.65	1.61	1.57	1.55	1.51	1.46	1.39	1.30
29											
	0.01	7.60	5.42	4.54	4.04	3.73	3.50	3.20	2.87	2.49	2.03
	0.05	4.18	3.33	2.93	2.70	2.54	2.43	2.28	2.10	1.90	1.64
	0.10	2.89	2.50	2.28	2.15	2.06	1.99	1.89	1.78	1.65	1.47
	0.20	1.72	1.70	1.65	1.60	1.57	1.54	1.50	1.45	1.39	1.29
30											
	0.01	7.56	5.39	4.51	4.02	3.70	3.47	3.17	2.84	2.47	2.01
	0.05	4.17	3.32	2.92	2.69	2.53	2.42	2.27	2.09	1.89	1.62
	0.10	2.88	2.49	2.28	2.14	2.05	1.98	1.88	1.77	1.64	1.46
	0.20	1.72	1.70	1.64	1.60	1.57	1.54	1.50	1.45	1.38	1.28
40											
	0.01	7.31	5.18	4.31	3.83	3.51	3.29	2.99	2.66	2.29	1.80
	0.05	4.08	3.23	2.84	2.61	2.45	2.34	2.18	2.00	1.79	1.51
	0.10	2.84	2.44	2.23	2.09	2.00	1.93	1.83	1.71	1.57	1.38
	0.20	1.70	1.68	1.62	1.57	1.54	1.51	1.47	1.41	1.34	1.24
60											
	0.01	7.08	4.98	4.13	3.65	3.34	3.12	2.82	2.50	2.12	1.60
	0.05	4.00	3.15	2.76	2.52	2.37	2.25	2.10	1.92	1.70	1.39
	0.10	2.79	2.39	2.18	2.04	1.95	1.87	1.77	1.66	1.51	1.29
	0.20	1.68	1.65	1.59	1.55	1.51	1.48	1.44	1.38	1.31	1.18
120											
	0.01	6.85	4.79	3.95	3.48	3.17	2.96	2.66	2.34	1.95	1.38
	0.05	3.92	3.07	2.68	2.45	2.29	2.17	2.02	1.83	1.61	1.25
	0.10	2.75	2.35	2.13	1.99	1.90	1.82	1.72	1.60	1.45	1.19
	0.20	1.66	1.63	1.57	1.52	1.48	1.45	1.41	1.35	1.27	1.12
∞											
	0.01	6.64	4.60	3.78	3.32	3.02	2.80	2.51	2.18	1.79	1.00
	0.05	3.84	2.99	2.60	2.37	2.21	2.09	1.94	1.75	1.52	1.00
	0.10	2.71	2.30	2.08	1.94	1.85	1.77	1.67	1.55	1.38	1.00
	0.20	1.64	1.61	1.55	1.50	1.46	1.43	1.38	1.32	1.23	1.00

* Table 9.10 is abridged from Table V of Fisher and Yates: *Statistical Tables of Biological, Agricultural, and Medical Research*, published by Oliver and Boyd Ltd., Edinburgh, by permission of the author and publishers.

within. Assume a 1 per cent level test. What is the value of F that you must obtain in order to reject your null hypothesis? To answer this question, enter the column labeled "5" and read down until you find the rows for 54 df. A complication arises—there is no row for 54 df. In this case you must interpolate. 54 df falls between the tabled values of 40 df and 60 df. If you had had 40 df for your within groups, then you would have needed a computed F of 3.51 in order to reject your null hypothesis; similarly if you had had 60 df, you required an F of 3.34. By linearly interpolating we find that an F of 3.39 is required for significance at the 1 per cent level. Try some additional problems on yourself.

Now, if you had conducted the above experiment you might feel quite happy with yourself—you have succeeded in rejecting the null hypothesis. But wait a moment. What does your null hypothesis say? It says that there is (at least) a difference among your groups. But where does the difference lie? Is it between Groups 1 and 2, between Groups 1 and 3, between Groups 2 and 3, or are two, or all, of these differences significant? The conventional answer to this question is to run t-tests between the groups as indicated above; this involves running three t-tests. Using the data in Table 9.8, compute the values of t; you will find them to be:

Between Groups 1 and 2: $t_{12} = 2.08$
Between Groups 1 and 3: $t_{13} = 3.57$
Between Groups 2 and 3: $t_{23} = 2.10$

Assuming a significance level of 0.05, we find that only the t between Groups 1 and 3 is significant, and our question as to where among the three groups the significant difference lies is answered.[10]

To briefly summarize the general approach of this section, we may say that on the assumption that only comparisons between pairs are to be made, and on the assumption that all possible pairs are to be compared, we first run an F test. If the value of this "over-all" F is significant, we then conduct all possible t tests in order to ascertain which specific groups are significantly different. If, however, the overall F is not significant, then we conclude that there are no significant differences between the various pairs of groups, thereby not running additional t tests.

Let us briefly comment on a different approach that you might wish to take in an experiment. Suppose that you are not really interested in making all possible comparisons between your groups. For instance, suppose that you conduct a three group experiment, and that your

[10] Obviously we have now tested three additional null hypotheses, set up for our three t tests, e.g., there is no difference between the means of Groups 1 and 2.

empirical hypothesis suggests that Groups 1 and 2 should differ and that Groups 1 and 3 should also differ; but in this event the comparison between Groups 2 and 3 is rather uninteresting to you—your hypothesis says nothing about this comparison. In this event you need not conduct your overall F test—you go directly into your t test analysis, computing the two indicated values of t.[11]

Appendix to Chapter 9

To illustrate the inappropriateness of applying the t-test to the multi-randomized groups design, suppose that we run a two-groups experiment on rats. We set our significance level at 0.05. Recall that, assuming that the null hypothesis is true, this significance level means that if we obtain a t that has a P of 0.05, the odds are five in one hundred that a t of this size or larger could have occurred by chance. Since this is a very unlikely occurrence, we reason that the t did not occur by chance. Rather, we prefer to conclude that the two groups are "really" different as measured by the dependent variable. We thus reject our null hypothesis and conclude that variation of the independent variable was effective in producing the difference between our two groups. Now, after completing the above work, say that we conduct a new two-groups experiment, for example one on schizophrenics. Note that the two experiments are independent of each other. In the experiment on schizophrenics we also set our significance level at 0.05, and follow the same procedure as before. Again our significance level means that the odds are five in a hundred that a t of the corresponding size could have occurred by chance.

But let us ask a question. Given a significance level of 0.05 in each of the two experiments, what are the odds that by chance the t in one, the other, or both experiments will be significant? Before you reach a hasty conclusion, let us caution you that the probability is *not* 0.05. Rather, the joint probability could be shown to be 0.0975.[12] That is, the odds of obtaining a t significant at the 0.05 level in either or both experiments are 975 out of 10,000. And this is certainly different from 0.05.

To illustrate, we might develop an analogy: what is the probability of obtaining a head in two tosses of a coin? On the first toss it is one

[11] Incidentally we might observe that this is a legitimate analysis, not subject to the difficulties indicated in the appendix to this chapter. For you are allowed to make two independent comparisons, which you do in this example, thereby not disturbing your "real" value of P.
[12] By the following formula: $P_j = 1 - (1 - \alpha)^k$ where P_j is the joint probability, α is the significance level, and k is the number of independent experiments. For instance in this case $\alpha = .05$, $k = 2$. Therefore, $P_j = 1 - (1 - 0.05)^2 = 0.0975$.

in two, and on the second toss it is one in two. But the probability of obtaining two heads on two successive tosses (before your first toss) is $\frac{1}{2} \times \frac{1}{2} = \frac{1}{4}$. To develop the analogy further, the probability of obtaining a head on the first toss, or on the second toss, or on both tosses (again, computed before *any* tosses) is $P = 0.75$.

In a different situation, suppose that we conduct a three-groups experiment. In this case there are three t tests in which we would probably be interested: a t between Groups I and II, between Groups I and III, and between Groups II and III. Assume that we set a significance level of 0.05 *for each* t. If the first t test yields a value significant beyond the 0.05 level, we reject the null hypothesis. And likewise for the other two t tests. But what are the odds of obtaining a significant t when we consider all three t tests? That is, what are the odds of obtaining a significant t in at least one of the following situations:

First: Between Groups I and II
or Second: Between Groups I and III
or Third: Between Groups II and III
or Fourth: Between Groups I and II and also between Groups I and III
or Fifth: Between Groups I and II and also between Groups II and III
or Sixth: Between Groups I and III and also between Groups II and III
or Seventh: Between Groups I and II and also between Groups II and III and also between Groups I and III.

The answer to this question is more complex than before. The reason for this is that these t tests are *not* independent. For example, the computed value of t between Groups I and II would be related to the computed value of t between Groups I and III because Group I occurs in both t tests. Similarly, t tests between Groups II and III and between Groups I and III would not be independent. And lacking independence in the t tests, it would be difficult to say just what the joint or over-all significance level is (as we were able to say in the previous case that it was 0.0975). About the best we can say for the general case is that the significance level for all possible t tests is less than that which would obtain if the t tests were independent. But this is not much help. The moral, then, is that the running of all possible t tests (between pairs of means) is not a satisfactory technique of statistical analysis—it simply does not provide a reasonable level of significance when all possible t's are considered. But this is not the worst of it. If, for instance, we had seven groups in our experiment, we would have to run 21 t tests in order to consider all possible combinations between pairs of means. Although we would not shirk the great work involved in running all these tests if it were necessary, we still have a limited amount of energy and would prefer to expend it in ways other than running a great many t tests, if possible.

One appropriate solution might be to run fewer t tests. Assuming that we are only interested in t tests between pairs of means, we could select those t tests that are independent and run them. Following the seven-groups example, however, it could be shown that only three such t tests *between pairs* would be independent. And we are usually interested in comparing more than three pairs of groups in such an experiment.

The importance of this discussion is that we have demonstrated our objections to the more frequently used procedure of analyzing the multi-randomized-groups design, that of an analysis of variance followed by all possible t tests. These criticisms are not directed toward the analysis of variance phase, for that by itself is perfectly legitimate. Thus you may conduct your analysis of variance and run your F test. If it is significant, then you know that there is a significant difference between at least two of your groups—but that is all that the F test tells you, for you do not know where the difference lies.

Duncan's Range Test seems considerably more appropriate for it: (1) allows us to make all possible comparisons between pairs of our groups, just as the 21 t tests in the previous example would; (2) is considerably less work than running a number of t tests; and (3) provides a more reasonable level of significance for all possible t tests, considered jointly. It is to be expected that in the foreseeable future Duncan's Range Test will be much more widely used than at the present time unless, of course, an even more desirable statistical test becomes available. As we indicated earlier, considerable work in this area is being conducted, and we can well expect important advances to be made (e.g., Duncan, 1959). Approaches other than those that we have presented are certainly available (cf. Ryan, 1959; Gaito, 1959).

Summary of the Computation of Duncan's Range Test for a Randomized-Groups Design with More than Two Groups

Assume that the following dependent variable scores have been obtained for four groups of subjects.

Group 1	Group 2	Group 3	Group 4
1	2	8	7
1	3	8	8
3	4	9	9
5	5	10	9
5	6	11	10
6	6	12	11
7	6	12	11

1. First we wish to compute ΣX, ΣX^2, n, and $\overline{X}$ values for each group.

	Group 1	Group 2	Group 3	Group 4
ΣX:	28	32	70	65
ΣX^2:	146	162	718	617
n:	7	7	7	7
$\bar{X}$:	4.00	4.57	10.0	9.29

2. Compute the sum of squares (SS) for each group. These values are determined by substituting in equation 9.1 and performing the indicated operations.

$$SS = \Sigma X^2 - (\Sigma X)^2/n$$
$$SS_1 = 146 - (28)^2/7 = 34.00$$
$$SS_2 = 162 - (32)^2/7 = 15.71$$
$$SS_3 = 718 - (70)^2/7 = 18.00$$
$$SS_4 = 617 - (65)^2/7 = 13.43$$

3. Using equation 9.5, compute the square root of the error variance. Substituting the above values and performing the appropriate operations, we find that

$$s_e = \sqrt{\frac{SS_1 + SS_2 + SS_3 + SS_4}{4(n-1)}} = 1.84$$

4. Determine the degrees of freedom for Duncan's Range Test, where

$$df = N - r = 28 - 4 = 24$$

5. Assuming a one per cent level test, enter Table 9.5 to determine the appropriate values of r_p. Since we have four means in the present example, we need to enter Table 9.5 at the columns labeled 4, 3, and 2. With 24 df we find the values of r_p to be

Number of Groups

	2	3	4
r_p	3.96	4.14	4.24

6. Compute the least significant ranges (R_p) for comparisons between two groups, among three groups, and among four groups. The equation (equation 9.4) is

$$R_p = (s_e)(r_p)\sqrt{1/n}$$

Making the appropriate substitutions to determine R_p for two groups (i.e., R_2), and performing the indicated operations,

$$R_2 = (1.84)(3.96)\sqrt{1/7} = 2.76$$

Similar substitutions and performance of the operations results in the values of R_p for three and for four groups:

$$R_3 = (1.84)(4.14)\sqrt{1/7} = 2.90$$
$$R_4 = (1.84)(4.24)\sqrt{1/7} = 2.96$$

The computed values of R_p may now be summarized:

Number of Groups

	2	3	4
R_p:	2.76	2.90	2.96

7. The next step is to rank the means of the groups from lowest to highest.

	Group 1	Group 2	Group 4	Group 3
$\bar{X}$:	4.00	4.57	9.29	10.00

8. We now test for significant differences among the various pairs of means. Starting with the highest (Group 3) and the lowest (Group 1), we can see that the difference between their means is 6.00. Determining the appropriate value of R_p for the comparison (that for four groups, hence $R_4 = 2.96$) we compare the mean difference with the value of R_p. In this case 6.00 exceeds 2.96. Therefore, Groups 1 and 3 differ significantly. The next comparison is between the highest group (Group 3) and the second from the lowest group (Group 2). The difference between these pairs of means is 5.43. The value of R_p for comparing three groups is 2.90. Since 5.43 exceeds 2.90, groups 2 and 3 differ significantly. The next comparison is between the highest and the next to highest means. The mean difference is 0.71. This is a two-group comparison; hence $R_p = 2.76$. Since the difference between the means of Groups 3 and 4 (.71) does not exceed 2.76, these two groups do not differ significantly. We now test for a significant difference between the next to highest mean and the lowest mean. The mean difference is 5.29 and $R_3 = 2.90$. Therefore, Groups 4 and 1 differ significantly. In the test between Groups 4 and 2 the mean difference is 4.72 and $R_2 = 2.76$. Therefore Groups 4 and 2 differ significantly. The final comparison is between Groups 1 and 2. Their mean difference is .57. Since $R_2 = 2.76$, these two groups do not differ significantly.

9. We now summarize the results of our tests of significance. We found, in our example, that significant differences did not exist between Groups 1 and 2, nor between Groups 3 and 4. All other differences were significant. These findings are summarized as follows:

Group 1	Group 2	Group 4	Group 3
4.00	4.57	9.29	10.00

Problems

1. An experimenter was interested in assessing the relative sociability scores of different majors in his college. He selected a random sample of students who were majoring in English, Art and Chemistry and ad-

ministered a standardized test of Sociability. Assuming a 1 per cent level of significance, did the three groups differ on this measure? Can it be said that all three groups are significantly different from each other?

Sociability Scores

English Majors	Art Majors	Chemistry Majors
1, 2, 2, 2, 3, 3	5, 5, 5, 6, 6, 6	9, 9, 9, 9, 10, 10

2. A Physical Education professor is interested in the effect of practice on the frequency of making goals in hockey. After consulting a psychologist he designs the following experiment. Four groups are formed such that Group 1 received the most practice, Group 2 the secondmost practice, Group 3 the thirdmost practice, and Group 4 the least amount of practice. Dependent variable scores represent the number of goals made by each subject during a test period. Determine which groups are significantly different. (Significance level is 0.05)

Group

1	2	3	4
5	1	5	10
7	4	0	9
9	0	2	6
7	0	1	5
6	8	1	8
5	3	4	9
9	2	3	8
2	1	0	2

3. An experimenter was interested in testing the hypothesis that the greater the hunger drive, the more correct choices a rat would make in a certain number of runs in a maze. He formed five groups of rats such that Group 1 had zero hours of food deprivation, Group 2 had 12 hours, Group 3 had 24 hours of food deprivation, Group 4 had 36 hours of food deprivation, and Group 5 had 48 hours of food deprivation. Setting a 5 per cent level of significance, was his hypothesis confirmed?

Number of Correct Choices

Group 1	Group 2	Group 3	Group 4	Group 5
0	1	0	3	4
0	1	1	3	5
1	3	1	4	6
2	3	2	5	7
3	4	4	6	7
3	4	4	7	8
4	4	5	7	9
5	4	5	8	10
6	5	7	9	11
7	6	8	10	12
7	7	9	11	14

4. An experiment is conducted to determine which of three methods of teaching Spanish is superior. Assuming that the experiment has been adequately conducted, that the 5 per cent level of significance has been set, and that the higher the test score the better the performance after training on the three methods, which method is to be preferred?

Method A Subjects	Method B Subjects	Method C Subjects
55	46	45
52	40	41
50	35	37
48	32	36
47	31	30
46	28	25
40	25	24
35	22	21
	21	21
	19	20
		19
		18
		17

IO

Experimental Design: The Factorial Design

All of the preceding designs are appropriate to the investigation of a single independent variable. If the independent variable is varied in two ways, one of the two-groups designs is used. If the independent variable is varied in more than two ways, the multi-group design is used. It is possible, however, to study more than one independent variable in a single experiment. One possible design for studying two or more independent variables in a single experiment is the *factorial design*. A factorial design is one where all possible combinations of the selected values of each of the independent variables are used. To illustrate a simple factorial design, let us say that we are interested in evaluating the effects of two independent variables on performance: amount of fatigue and meaningfulness of material. Assume that we wish to select two values of each independent variable for study. Amount of fatigue might be little and considerable (or more precisely, the amount of fatigue that occurs one hour after sleep and twenty-four hours after sleep). Meaningfulness of material might be prose and nonsense syllables (where prose is highly meaningful but nonsense syllables are not). Variation of these two independent variables might be diagrammed as in Figure 10.1. But since we wish to employ a factorial design, all possible combinations.

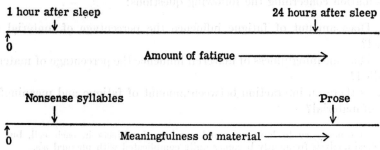

Figure 10.1. Variation of two independent variables, each in two ways.

of the values of the independent variables are to be used, as indicated in Figure 10.2.

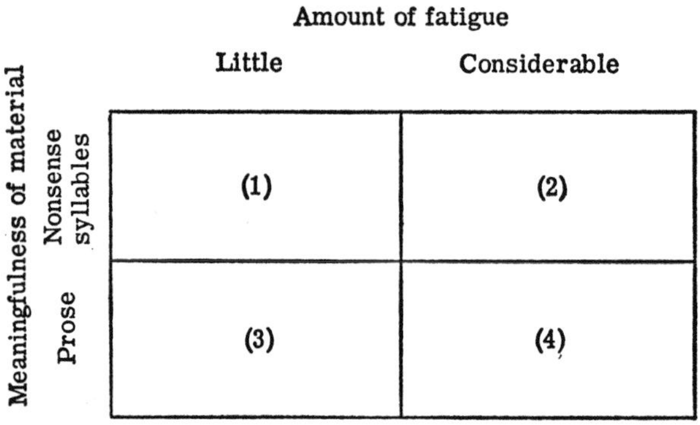

Amount of fatigue

Figure 10.2. Diagram of a factorial design.

Figure 10.2 shows that there are four possible combinations of the values of the independent variables. Each possible combination is represented by a square, a *cell:* (1) little fatigue and nonsense syllables; (2) considerable fatigue and nonsense syllables; (3) little fatigue and prose; and (4) considerable fatigue and prose. With four experimental conditions it follows that we shall have four groups in the experiment. Therefore, we shall assign an equal number of whatever sample of subjects we have drawn to the four conditions.[1]

The experiment would then be conducted as follows: Group 1 would learn a list of nonsense syllables one hour after they have awakened from sleep; Group 2 would learn the same list of nonsense syllables twenty-four hours after they have slept; Group 3 would learn a certain passage of prose one hour after sleep; and Group 4 would learn the same passage twenty-four hours after sleep. Our dependent variable might be the percentage of material recalled at the end of a certain period of training. A statistical analysis of these data would then provide information concerning the following questions:

1. Does amount of fatigue influence the percentage of material recalled?

2. Does meaningfulness of material influence the percentage of material recalled?

3. Is there an interaction between amount of fatigue and meaningfulness of material?

[1] It is not necessary to have an equal number of subjects in each cell, but the statistical analysis frequently becomes quite complicated with unequal n's.

Illustration of possible answers to the first two questions is straightforward, but the third will require a little more consideration. Assume that we have forty subjects available and randomly assign ten to each group. Further assume that the dependent variable scores obtained for each group are as follows (Table 10.1).

Table 10.1

Fictitious Dependent Variable Scores for the Four Groups
That Compose the Factorial Design of Fig. 10.2

Group

1	2	3	4
(Little fatigue–Nonsense Syllables)	(Considerable fatigue–Nonsense Syllables)	(Little fatigue–Prose)	(Considerable fatigue–Prose)
90	88	100	99
65	75	88	82
42	62	95	84
75	70	97	97
68	35	92	96
76	76	98	92
65	63	67	98
70	71	89	94
72	69	94	89
68	70	97	97
n: 10	10	10	10
$\bar{X}$: 69.1	67.9	91.7	92.8
$\Sigma X = 691.$	$= 679.$	$= 917.$	$= 928.$
$\Sigma X^2 = 49,047$	$= 47,785$	$= 84,901$	$= 86,440$

Now let us place the means for the four groups in their appropriate cells (Figure 10.3).

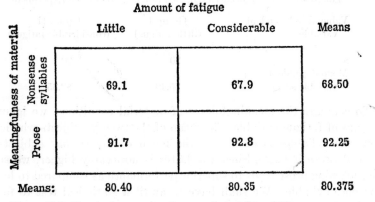

Figure 10.3. Showing means for the experimental conditions.

Let us study the effect of variation of amount of fatigue (question 1). For this purpose we shall ignore the meaningfulness-of-material variable. That is, we have ten subjects who learned nonsense syllables under little fatigue, and ten other subjects who learned prose under little fatigue. Ignoring the fact that ten subjects learned nonsense syllables and ten learned prose, we have twenty subjects who learned under the condition of little fatigue. Similarly, we have twenty subjects who learned under a condition of considerable fatigue. We, therefore, have two groups of subjects who, as a whole, were treated identically except with regard to amount of fatigue. For the amount of fatigue comparison it is irrelevant that half of each group learned different types of material —the material variable is balanced out. To make our comparison we need merely compute the mean for the twenty low-fatigue subjects and for the twenty high-fatigue subjects. To do this we have computed the mean of the means for the two conditions. (This is possible because the n's for each mean are equal.) That is, the mean of 69.1 and 91.7 is 80.4 and similarly for the considerable fatigue condition, as shown in Figure 10.3. Since the two means (80.40 and 80.35) are almost identical, it is clear that amount of fatigue did not influence the dependent variable.

Students who find it difficult to ignore the meaningfulness-of-material variable when considering the amount-of-fatigue variable should look at the factorial design as if they are conducting only one experiment, and varying only the amount of fatigue. In this case the meaningfulness-of-material variable can be temporarily considered as an extraneous variable whose effect is balanced out. Thus, the two-groups design would look like that indicated in Table 10.2.

Table 10.2

Looking at the Factorial Design as a Single Two Groups Experiment

Value of independent variable	Group I (little fatigue)	Group II (considerable fatigue)
n	20	20
Mean dependent variable score	80.40	80.35

To compare the meaningfulness-of-material conditions, we ignore the amount of fatigue variable. The mean of the twenty subjects who learned nonsense syllables is 68.50 and the mean of the twenty subjects who learned prose is 92.25. Since the latter is noticeably higher, we suspect that learning prose leads to superior performance as compared to learning nonsense syllables. We shall have to await a statistical test to find out if this difference is significant.

Now that we have preliminary answers to the first two questions, let us turn to the third: is there an interaction between the two variables? Interaction is one of the most important concepts discussed in this book. If you adequately understand it you will have ample opportunity to apply it in a wide variety of situations—it will shed light on a large number of problems and increase your understanding of behavior considerably.

First, let us approach the concept of interaction from an overly simplified point of view. Assume the problem is of the following sort: Is it more efficient (timewise) for a man who is dressing to put his shirt or his trousers on first? At first glance it might seem that a suitable empirical test would yield one of two answers: (1) shirt first or (2) trousers first. However, in addition to these possibilities there is a third answer—(3) it depends. Now "it depends" embodies the basic notion of interaction. Suppose a finer analysis of the data indicates what "it depends" on. We may find that it is more efficient for tall men to put their trousers on first, but for short men to put their shirt on first. In this case we may say that our answer depends on the body build of the man who is dressing. Or to put it in terms of an interaction, we may say that there is an interaction between putting trousers or shirt on first with body build. This is the basic notion of interaction. Let us take another example from everyday life before we consider the concept in a more precise manner.

The author once had to obtain the support of a senior officer in the Army to conduct an experiment. In order to control certain variables (e.g., the effect of the company commander) it was decided to use only one company. There were four methods of learning to be studied so it was planned to divide the company into four groups. Each group (a platoon) would then learn by a different method. The officer, however, objected to this design. He said that "we always train our men as a whole company. You are going to train the men in platoon sizes. Therefore, whatever results you obtain with regard to platoon-size training units may not be applicable to what we normally do with company-size units." The author had to admit this point, and it is quite a sophisticated one. It is possible that the results for platoons might be different than the results for companies—that there is an *interaction* between size of training unit and the type of method used. In other words, one method might be superior if used with platoons, but another if used with companies. Actually, previous evidence suggested that such an interaction is highly unlikely in this situation, so the author didn't worry about it—he only left a slightly distressed senior officer.

An interaction exists between two independent variables if the dependent variable value that results from one independent variable is determined by the specific value assumed by the other independent variable. To illustrate, assume that there is an interaction between the two variables of fatigue and meaningfulness of material. Such an interaction would be stated as follows: the results with regard to amount of fatigue depend upon the type of material learned. Or more precisely: the superiority of one amount of fatigue depends upon whether nonsense syllables or prose was learned. In Figure 10.4, we have plotted some fictitious de-

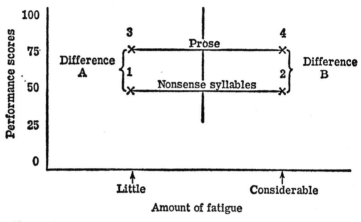

Figure 10.4. Illustration of lack of interaction between amount of fatigue and meaningfulness of material.

pendent variable scores along the vertical axis. On the horizontal axis we have represented two values of the amount-of-fatigue variable. The data points represent the means of the four conditions: point number one is the mean for the little fatigue–nonsense syllable condition; two is for the considerable fatigue–nonsense syllable condition; three the little fatigue–prose condition; and four the considerable fatigue–prose condition. The line that connects points one and two represents the performance of the subjects who learned nonsense syllables, half under little fatigue and half under considerable fatigue. The line through points three and four represents the performance of the subjects who learned prose. Referring to Figure 10.4, let us ask what the effects of the independent variables were. First, variation of amount of fatigue leads to negligible variation of the dependent variable, for both lines are approximately horizontal. Second, the subjects who learned prose performed better than the subjects who learned nonsense syllables (the "prose" line is higher than the "nonsense syllable" line). And third, the difference in performance between the nonsense-syllable and prose subjects who had little fatigue (Difference A) is approximately the same as the difference between the nonsense syllable and prose subjects who had considerable fatigue (Difference B). It may be seen that the effect of the type of material learned is independent of the effect of amount of fatigue—no interaction between these two variables exists. Put another way: if the lines drawn in Figure 10.4 are approximately parallel (i.e., if Difference A is approximately the same as Difference B), it is likely that no interaction exists between the variables.[2] However, if the lines are clearly not parallel (i.e., if Difference A is not equal to Difference B), an interaction is present.

Another way of illustrating the same point is to compute the differences between the means of the groups. The necessary differences are shown in Figure 10.5. The difference between the nonsense syllable and

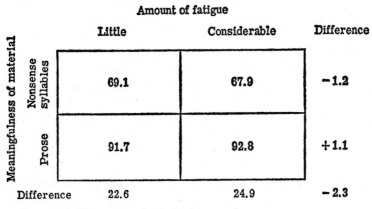

Figure 10.5. Illustration of a lack of interaction.

the prose groups for little fatigue is 22.6, while for considerable fatigue it is 24.9. Since these differences do not differ to any great extent, we suspect that there is no interaction present. The same conclusion could be reached by comparing the differences in the other direction, i.e., since 1.2 and −1.1 are approximately the same, no interaction exists. Incidentally, the 22.6 is Difference A of Figure 10.4, while 24.9 is Difference B. Clearly if these differences are about the same, the lines will be approximately parallel.

At this point you may be disappointed that the data did not yield an interaction. This can easily be arranged by assuming for the moment that the data came out as indicated in Figure 10.6. In this case our lines would look like those in Figure 10.7.

[2] Of course, as before, we are talking about sample values and not about population values. Thus, while this statement is true for sample values it is not true for population (true) values. Therefore, if the lines for the population values are even slightly non-parallel, there is an interaction.

Amount of fatigue

		Little	Considerable	
Meaningfulness of material	Nonsense syllables	69.1	90.0	79.55
	Prose	91.7	80.0	85.85
		80.40	85.00	84.20

Figure 10.6. New fictitious means designed to show interaction.

Now we note that the lines are not parallel—in fact they cross each other. Hence we may make the following statements: prose is superior with little fatigue, but nonsense syllables is superior with considerable fatigue; or the logically equivalent statement that little fatigue is superior with prose, but considerable fatigue is superior with nonsense syllables. Put in other words: the difference between nonsense syllables and prose depends on amount of fatigue, or equally, the difference between amounts of fatigue depend on meaningfulness of material.

This discussion should clarify the meaning of "interaction." This is a rather difficult concept, however, and the examples in the remainder of the chapter should help to illuminate it further.

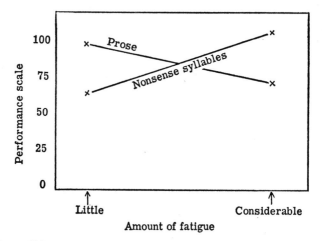

Figure 10.7. Illustration of a possible interaction.

Summary

When selected values of more than two independent variables are studied in all possible combinations, a factorial design is used. We have illustrated the factorial design by using two independent variables with two values of each. In this case subjects are randomly assigned to the four experimental conditions. Analysis of the dependent variable data yields information on (1) the influence of each independent variable on the dependent variable, and (2) the interaction between the two independent variables.

Types of Factorial Designs

Factorial Designs with Two Independent Variables

The 2 × 2 Factorial Design. The type of factorial design that we have discussed is referred to as the 2 × 2 factorial design. In this design we study the effect of two independent variables each varied in two ways. The number of numbers used in the label indicates the number of independent variables studied in the experiment. And the size of the numbers indicates the number of values of the independent variables. Since the 2 × 2 design has two numbers (2 and 2) we can tell immediately that there are two independent variables. And since the actual numbers are both 2, we know that each independent variable assumed two values. From "2 × 2" we can also tell how many experimental conditions there are—2 multiplied by 2 is 4.

The 2 × 3 Factorial Design. The 2 × 3 factorial design is one in which two independent variables are studied, one being varied in two ways, while the second assumes three values. Let us say that the preceding example was modified to form a 2 × 3 (it could as well be written

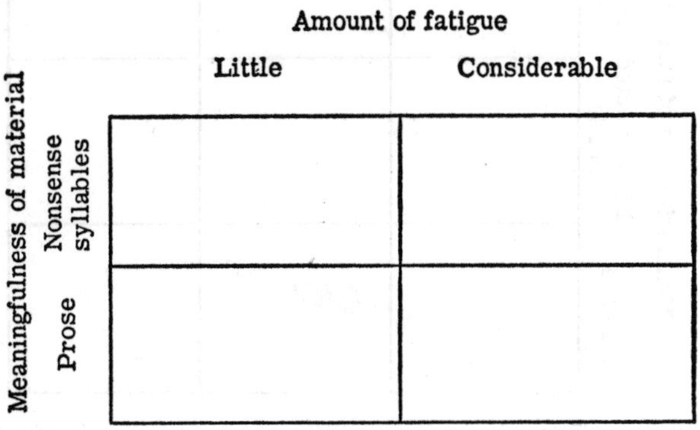

Figure 10.8. A 2 × 2 factorial design.

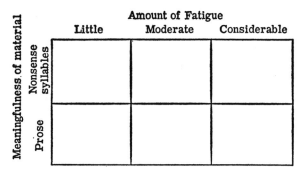

Figure 10.9. Extension of the design of Figure 10.8 to show an example of a 2 × 3 factorial design.

3 × 2) design. The amount of fatigue is varied in three ways: little, moderate, and considerable. Maintaining the same values for meaningfulness of material a diagram of this 2 × 3 design would be as in Figure 10.8. Now we have six conditions and therefore would randomly assign our subjects to six groups, instead of four. The experimental procedure would be the same except for the addition of the two moderate fatigue groups.

The 3 × 3 Factorial Design. This design is one in which we investigate two independent variables, each varied in three ways. We have nine experimental conditions, if we add, say, a medium condition of meaningfulness of material, such as disconnected English words, to the 3 × 2 design (Figure 10.10):

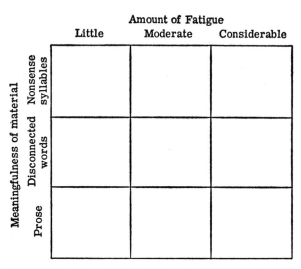

Figure 10.10. Extension of Figure 10.9 to show an example of a 3 × 3 factorial design.

The K × L Factorial Design. Each independent variable may be varied in any number of ways. The generalized factorial design for two independent variables may be labeled the K × L factorial design, where K stands for the first independent variable and its value indicates the number of ways in which it is varied, and L similarly denotes the second independent variable. K and L might then assume any value. If one independent variable is varied in two ways and the other in four ways, we would have a 2 × 4 design. If one independent variable is varied in two ways and the second in six ways, we would have a 2 × 6 design. If three values are assumed by one independent variable and five by the other we would have a 3 × 5 design, and so forth. A 4 × 7 design would look like Figure 10.11.

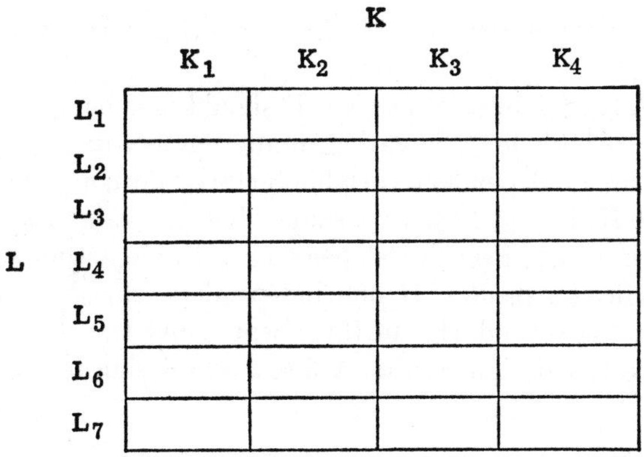

K

	K_1	K_2	K_3	K_4
L_1				
L_2				
L_3				
L_4				
L_5				
L_6				
L_7				

Figure 10.11. Diagram of a 4 × 7 factorial design.

Factorial Designs with More than Two Independent Variables

The 2 × 2 × 2 Factorial Design. The previous factorial designs have concerned two independent variables. However, in principle the number of variables that can be studied is unlimited. The 2 × 2 × 2 design is the simplest factorial for studying three independent variables. Following our preceding rule this label implies that each of three independent variables is varied in two ways. It also follows that there are eight experimental conditions. An illustration of what happens when, in addition to varying fatigue and meaningfulness of material in two ways we also vary in two ways a third variable, situational stress, is presented in Figure 10.12.

The cells have been labeled with letters. As before, subjects would be randomly assigned to the cells. The subjects in Cell A would then learn under a little-fatigue, little-stress, prose condition. Those in Cell B

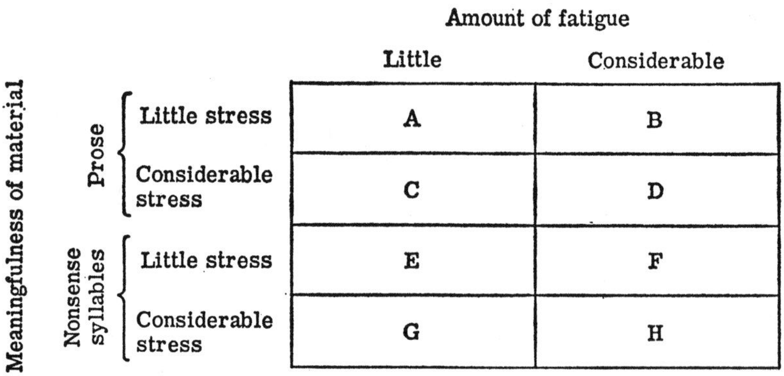

Amount of fatigue

			Little	Considerable
	Prose	Little stress	A	B
		Considerable stress	C	D
	Nonsense syllables	Little stress	E	F
		Considerable stress	G	H

Meaningfulness of material

Figure 10.12. Diagram of a $2 \times 2 \times 2$ factorial design.

would learn under a considerable-fatigue, little-stress, prose condition, and so on.

The K × L × M Factorial Design. It should now be apparent that any independent variable may be varied in any number of ways. The general case for the three independent variable factorial design is the K × L × M design, where K, L, and M may assume whatever value the experimenter desires. For instance, if each independent variable assumes three values a $3 \times 3 \times 3$ design results. If one independent variable (K) is varied in two ways, the second (L) in three ways, and the third (M) in four ways, a $2 \times 3 \times 4$ design results. A $3 \times 5 \times 5$ design is shown in Figure 10.13.

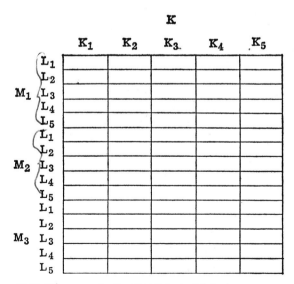

Figure 10.13. Diagram of a $3 \times 5 \times 5$ factorial design.

Statistical Analysis of Factorial Designs

We have compared the means for each of our experimental conditions and looked at the possible interaction in our example, but this has provided only tentative answers; firmer answers await the application of statistical tests to our data. Where we obtained a sizeable difference between the subjects who learned by prose and by nonsense syllables, for example, we said that we had to find out if the apparently sizeable difference in means was significant. The statistical analysis that is most frequently applied to the factorial design is *analysis of variance*, the rudiments of which were presented in Chapter **9**. We shall limit our discussion to the **2 × 2** factorial design.

The first step in conducting an analysis of variance for the factorial design follows very closely that for any number of groups, as previously discussed. That is, we wish to compute the total *SS* and partition it into two major components, the between *SS* and the within *SS*. Consider the fictitious data in Table 10.3.

Table 10.3

Fictitious Scores for Four Groups that Compose a Factorial Design

	Group			
	1	*2*	*3*	*4*
	2	1	6	4
	3	2	7	5
	4	3	8	6
ΣX:	9	6	21	15
ΣX^2:	29	14	149	77
n:	3	3	3	3

To compute the total *SS*, we substitute the appropriate values from Table 10.3 in equation 9.9 which for four groups (always the case for the 2 × 2 design) is:

(10.1)

$$\text{Total } SS = (\Sigma X_1^2 + \Sigma X_2^2 + \Sigma X_3^2 + \Sigma X_4^2) - \frac{(\Sigma X_1 + \Sigma X_2 + \Sigma X_3 + \Sigma X_4)^2}{N}$$

$$= (29 + 14 + 149 + 77) - \frac{(9 + 6 + 21 + 15)^2}{12} = 52.25$$

Next, to compute the between *SS*, we substitute in equation 9.10, which for four groups is:

(10.2) Between $SS = \dfrac{(\Sigma X_1)^2}{n_1} + \dfrac{(\Sigma X_2)^2}{n_2} + \dfrac{(\Sigma X_3)^2}{n_3} + \dfrac{(\Sigma X_4)^2}{n_4}$

$$- \dfrac{(\Sigma X_1 + \Sigma X_2 + \Sigma X_3 + \Sigma X_4)^2}{N}$$

Between $SS = (9)^2/3 + (6)^2/3 + (21)^2/3 + (15)^2/3 - 216.75$

$= 44.25$

And, as before, the within SS may be obtained by subtraction (equation 9.11):

(10.3)   Within SS = Total SS − Between SS

$= 52.25 - 44.25 = 8.00$

This completes the initial stage of the analysis of variance for a 2×2 factorial design, for we have now illustrated the computation of the total SS, the between SS, and the within SS. As you can see, the initial stage of the statistical computation is the same as that for the initial stage for a randomized groups design. To follow through to its completion the analysis of variance for a 2×2 design, let us return to our previous example. The data for that experiment were presented on page 209, the necessary ingredients of which are reproduced in Table 10.4.

Table 10.4

Summary of the Necessary Ingredients for Analysis
of Variance Taken from Table 10.1

	1 (Little Fatigue– Nonsense Syllables)	2 (Considerable Fatigue– Nonsense Syllables)	3 (Little Fatigue– Prose)	4 (Considerable Fatigue– Prose)
n:	10	10	10	10
ΣX:	691	679	917	928
ΣX^2:	49,047	47,785	84,901	86,440

We seek to compute the following: total SS, between-groups SS, and within-group SS. Substituting in equation (10.1) we find the total SS to be:

Total $SS = 49,047 + 47,785 + 84,901 + 86,440$

$$- \dfrac{(691 + 679 + 917 + 928)^2}{40} = 9,767.38$$

The between-groups SS is determined by substituting the appropriate values in equation 10.2:

Between groups $SS = \dfrac{(691)^2}{10} + \dfrac{(679)^2}{10} + \dfrac{(917)^2}{10} + \dfrac{(928)^2}{10}$

$$- \frac{(691 + 679 + 917 + 928)^2}{40} = 5653.88$$

The within SS is determined by subtraction (equation 10.3):

$$\text{Within } SS = 9{,}767.38 - 5653.88 = 4113.50$$

The between-groups SS tells us something about how all groups differ. However, we are interested not in simultaneously comparing *all* groups, but only in certain comparisons. In a 2×2 factorial, of course we have two independent variables each varied in two ways. Hence we are interested in whether or not variation of each independent variable affects the dependent variable, and whether there is a significant interaction. The first step to perform is to compute the SS between groups for each independent variable. Using Figure 10.19 (p. 238) as a guide, we may write our formulas for computing the between groups SS for the specific comparisons.

The groups are as labeled in the cells. Thus to determine whether or not there is a significant difference between the two values of the first independent variable, we need to compute the SS between these two values. For this purpose we may use equation 10.4.

(10.4)

SS Between Amounts of First Independent Variable

$$= \frac{(\Sigma X_1 + \Sigma X_3)^2}{n_1 + n_3} + \frac{(\Sigma X_2 + \Sigma X_4)^2}{n_2 + n_4} - \frac{(\Sigma X_1 + \Sigma X_2 + \Sigma X_3 + \Sigma X_4)^2}{N}$$

For computing the SS between the conditions of the second independent variable we use equation 10.5:

(10.5)

SS Between Amounts of Second Independent Variable

$$= \frac{(\Sigma X_1 + \Sigma X_2)^2}{n_1 + n_2} + \frac{(\Sigma X_3 + \Sigma X_4)^2}{n_3 + n_4} - \frac{(\Sigma X_1 + \Sigma X_2 + \Sigma X_3 + \Sigma X_4)^2}{N}$$

Now, in our particular example we conduct tests to determine whether amount of fatigue (our first independent variable) influences the dependent variable, whether meaningfulness of material (our second independent variable) influences the dependent variable, and whether there is a significant interaction. First, to determine the effect of amount of fatigue we need to test the difference between the little-fatigue and the considerable-fatigue conditions. To make this test we shall ignore the meaningfulness-of-material variable in the design.

Making the appropriate substitutions in equation 10.4 we can compute the SS between the amount of fatigue conditions:

SS between fatigue conditions

$$= \frac{(691 + 917)^2}{10 + 10} + \frac{(679 + 928)^2}{10 + 10} - \frac{(691 + 679 + 917 + 928)^2}{40}$$

$$= \frac{(1608)^2}{20} + \frac{(1607)^2}{20} - \frac{(3215)^2}{40} = 0.02$$

This value will be used to answer the above question. However, we should answer all questions at once, rather than piecemeal, so let us hold it until we complete this stage of the inquiry. We have computed a sum of squares between all four groups (i.e., 5653.88). We shall refer to this sum of squares as the "over-all between-groups SS." And it can be separated into parts. We have computed one of these parts above, the sum of squares between the fatigue conditions (.02). There are two other parts: the sum of squares between the meaningfulness-of-material conditions, and the interaction. Let us compute the former, using equation 10.5.

Substituting the required values in equation 10.5 we determine that the "between" SS for the meaningfulness-of-material conditions is:

SS between meaningfulness conditions

$$= \frac{(691 + 679)^2}{10 + 10} + \frac{(917 + 928)^2}{10 + 10} - \frac{(691 + 679 + 917 + 928)^2}{40}$$

$$= 5640.63$$

The over-all between SS has three parts. We have directly computed the first two parts. Hence, the difference between the sum of the first two parts and the over-all between SS provides the third part, in this case that for the interaction:

(10.6)

Interaction SS = Over-all between SS − between SS for first variable (fatigue) − between SS for second variable (meaningfulness of material)

Recalling that the over-all between SS was 5653.88, the between SS for the fatigue conditions was 0.02 and the between SS for the meaningfulness-of-material conditions was 5640.63, we find that the SS for interaction is:

Interaction SS = 5653.88 − 0.02 − 5640.63 = 13.23

This completes the computation of the sums of squares. Let us add that these values should all be positive. If your computations yield a negative SS, check your work until you discover the error. There are only several minor matters to discuss before the analysis is completed. Before we continue, however, let us summarize our findings to this point in Table 10.5.

Table 10.5

Sums of Squares for the 2 × 2 Factorial Design

Source of Variation	Sum of Squares
Over-all between	(5653.88)
Between fatigue	0.02
Between material	5640.63
Interaction: material × fatigue	13.23
Within groups	4113.50
Total	9,767.38

We now must discuss how to determine various degrees of freedom for this application of the analysis-of-variance procedure. Repeating the equations in Chapter 9, for the major components:

(10.7)   Total $df = N - 1$
(10.8)   Between $df = r - 1$
(10.9)   Within $df = N - r$

In our example, $N = 40$, and r (number of groups) $= 4$. Hence, the total df is $40 - 1 = 39$, the over-all between df is $4 - 1 = 3$ (the over-all df is based on four separate groups or conditions), and the "within" df is $40 - 4 = 36$. The similarity between the manner in which we partition the total SS and the total df may also be continued for the over-all between SS and the over-all between df. The over-all between df is 3. Since we analyzed the over-all-between SS into three parts, we may do the same for the over-all-between df, one df for each part. Take the fatigue conditions first. Since we are temporarily ignoring the meaningfulness-of-material variable, we have only two conditions of fatigue to consider—or, if you will, two groups. Hence, the df for the between-fatigue conditions is based on $r = 2$. Substituting this value in equation 10.8, we see that the between-fatigue df is $2 - 1 = 1$. The same holds true for the meaningfulness-of-material conditions—there are two such conditions, hence $r = 2$ and the df for this source of variation is $2 - 1 = 1$.

Now for the interaction df. Note in Table 10.5 that the interaction is written as Material × Fatigue. We may, of course, have abbreviated the notation, as is frequently done, by using M × F. This is read "the interaction between material and fatigue." The "×" sign may be used as a mnemonic device for remembering how to compute the interaction df: multiply the number of degrees of freedom for the first variable by that for the second. Since both variables have one df, the interaction df is also one, i.e., $1 \times 1 = 1$. This accounts for all three df that are associated with the over-all between SS.[3] These findings are added to Table 10.5, forming Table 10.6.

Table 10.6

Sums of Squares and df for the 2 × 2 Factorial Design

Source of Variation	Sum of Squares	df
Over-all between	(5653.88)	(3)
Between fatigue	0.02	1
Between material	5640.63	1
Interaction: material × fatigue	13.23	1
Within groups	4113.50	36
Total	9,767.38	39

In the 2 × 2 factorial design there are four mean squares in which we are interested: between fatigue conditions, between material conditions, the interaction, and within groups. To compute the mean square for the between-fatigue source of variation, we divide that sum of squares by the corresponding df, $.02/1 = .02$. Similarly the within groups means square is computed as:

$$4113.50/36 = 114.26$$

These values are then added to our summary table of the analysis of variance, as we shall show shortly.

This completes the analysis of variance for the 2 × 2 design, at least in the usual form. We have analyzed the total variance into its various components. In particular, we have several sources of between variation to study and a term that represents the experimental error (the within-groups mean square). We said that the "between" components indicate the extent to which the various experimental conditions differ. For instance, if any given "between" component, such as that for the fatigue conditions, is sizeable, then that may be taken to indicate that amount of fatigue influences the dependent variable. Hence we need merely conduct the appropriate F tests to determine whether or not the various "between" components are significantly larger than chance. The first F for us to compute is that between the two conditions of fatigue.[4] To do

---

[3] If this is not clear to you, then you might merely remember that the df for the between SS in a 2 × 2 design are always the same as shown in Table 10.6. That is, the df for the over-all between SS is always 3, the df for the SS between each independent variable condition is 1 and for the interaction, 1.

[4] The factorial design offers us a good example of a point we made in the last chapter. That is that if we are specifically interested in certain questions, then there is no need to conduct an over-all F test. With this design we are exclusively interested

this we merely substitute the appropriate values in equation 9.17. Since the mean square between the fatigue conditions is .02 and the mean square within groups is 114.26, we divide the former by the latter,

$$F = 0.02/114.26 = 0.00$$

The F between the meaningfulness-of-material conditions is:

$$F = 5640.63/114.26 = 49.37$$

And the F for the interaction is:

$$F = 13.23/114.26 = 0.12$$

These values have been entered in Table 10.7.

Table 10.7

Complete Analysis of Variance of the Performance Scores

Source of Variation	Sum of Squares	df	Mean Square	F
Between fatigue	0.02	1	0.02	0.00
Between material	5640.63	1	5640.63	49.37
Interaction: $M \times S$	13.23	1	13.23	0.12
Within groups (error)	4113.50	36	114.26	
Total	9767.38	39		

Following the preceding discussion, let us observe that Table 10.7 is the final summary of our statistical analysis. This is the table that should be presented in the results section of an experimental write-up. We next assign a probability level to these values. That is, we need to determine the odds that the F's could have occurred by chance. As before we first set up a null hypothesis: there is no difference between our groups. However, we have three more precise null hypotheses in this type of design:

1. There is no difference between the two conditions of fatigue.
2. There is no difference between the two conditions of meaningfulness of material.
3. There is no interaction between the two independent variables.[5]

The general strategy is to determine the probability that each value of F could have occurred by chance. Assuming that we have set a significance level of 0.05 for each F test, we need merely determine the

in whether our two independent variables are effective, and whether there is an interaction. Hence we proceed directly to these questions without running an overall F test, although such may easily be conducted if you are so inclined.
[5] A more precise statement could be made; there is no difference in the means of the four groups after the cell means have been adjusted for row and column effects. However, such a statement probably will only be comprehensible to you after further work in statistics.

probability associated with each F. If that probability is 0.05 or less, we can reject the appropriate null hypothesis and conclude that the independent variable in question was effective in producing the result.[6] In the first null hypothesis, the value of F is 0.00. We already know that this F is not significantly large, since it is below a value of 1.0. Hence, we can conclude that the first null hypothesis cannot be rejected —variation of amount of fatigue does not affect the dependent variable. Now let us turn to the second null hypothesis, that for the meaningfulness of material conditions. Our obtained F is 49.37. We have one df for the numerator and 36 df for the denominator. We can determine that an F of 4.12 is required for significance at the 5 per cent level with one and 36 df (Table 9.10). Since our F of 49.37 exceeds this value, we may reject the second null hypothesis. The conclusion is that the two conditions of meaningfulness of material led to significantly different performance. And since the mean for the prose condition (92.25) is higher than for the nonsense syllable condition (68.5), we may conclude that learning prose leads to significantly superior performance when compared with learning nonsense syllables.

To test the interaction we note that the F is 0.12. As before, this F is considerably below 1.00. We can, therefore, conclude immediately that the interaction is not significant. A check on this may be made by noting that we also have one and 36 df for this source of variation. And we know that an F of 4.12 is required for significance at the 5 per cent level. Clearly 0.12 does not approach 4.12, and hence is not significant. The third null hypothesis is *not* rejected.

The preceding discussion for the statistical analysis of a factorial design has been rather lengthy because of its detailed nature. Now with this background it is possible for us to breeze through the next example. Suppose that we are interested in whether males or females can solve problems better, and also in whether differences in IQ are related to problem-solving ability. We might use the parlor game "Twenty Questions" as our test of problem-solving ability. The number of questions that a person needs to ask in order to get the answer would be our measure of the dependent variable. Assume that we have 30 male and 30 female subjects. We gave subjects an intelligence test and selected two extreme groups: high IQ and low IQ. Our 2 × 2 factorial design would then be as in Figure 10.14.

The null hypotheses may be stated as follows:

1. There is no difference between males and females.
2. There is no difference between high and low IQ's.
3. There is no interaction between sex and IQ.

[6] Of course, assuming adequate control procedures have been exercised.

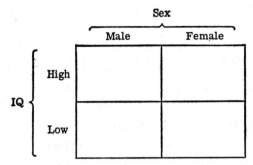

Figure 10.14. A 2 × 2 factorial design for investigation of the effects of sex and IQ on problem-solving.

We shall set a significance level of 0.01 as our criterion for rejecting the null hypotheses.

Now let us assume that we obtained the results tabulated in Table 10.8.

Table 10.8

Fictitious Results for the 2 × 2 Design of Figure 10.15

Group 1 High IQ-Male	Group 2 High IQ-Female	Group 3 Low IQ-Male	Group 4 Low IQ-Female
7	20	19	6
8	18	16	8
6	16	20	9
9	19	17	4
8	14	13	9
5	19	18	11
9	16	18	8
10	17	18	9
9	18	17	9
7	19	16	7
8	20	19	8
9	18	17	7
7	19	20	8
6	17	16	7
9	18	15	9
$\Sigma X = 117$	268	259	119
$\Sigma X^2 = 941$	4826	4523	981
$n = 15$	15	15	15

Our first step will be to compute the total SS by substituting the values in Table 10.8 in equation 10.1:

$$\text{Total } SS = 941 + 4826 + 4523 + 981 - \frac{(117 + 268 + 259 + 119)^2}{60}$$

$$= 1568.18$$

Now we shall compute the over-all between SS by appropriate substitutions in equation 10.2.

$$\text{Between } SS = \frac{(117)^2}{15} + \frac{(268)^2}{15} + \frac{(259)^2}{15} + \frac{(119)^2}{15}$$

$$- \frac{(117 + 268 + 259 + 119)^2}{60} = 1{,}414.18$$

The within SS is (see equation 10.3):

$$1568.18 - 1414.18 = 154.00$$

As our next step we shall analyze the over-all between SS into its three components: between sex, between IQ, and the S × I interaction. Considering sex first, we substitute the appropriate values in equation 10.4 and find that:

SS between sex conditions

$$= \frac{(117 + 259)^2}{15 + 15} + \frac{(268 + 119)^2}{15 + 15} - \frac{(117 + 268 + 259 + 119)^2}{60}$$

$$= 2.01$$

Substituting in equation 10.5 to compute the SS between the two conditions of IQ:

SS between IQ conditions

$$= \frac{(117 + 268)^2}{15 + 15} + \frac{(259 + 119)^2}{15 + 15} - \frac{(117 + 268 + 259 + 119)^2}{60}$$

$$= 0.82$$

The SS for the interaction component may now be seen to be:

$$1414.18 - 2.01 - 0.82 = 1411.35$$

The various df may now be determined.

$$\text{Total } (N - 1) = 60 - 1 = 59$$
$$\text{Over-all between } (r - 1) = 4 - 1 = 3$$
$$\text{Between sex} = 2 - 1 = 1$$
$$\text{Between IQ} = 2 - 1 = 1$$
$$\text{Interaction: S} \times \text{I} = 1 \times 1 = 1$$
$$\text{Within } (N - r) = 60 - 4 = 56$$

The mean square and the F's have been computed and placed in the summary table.

Interpreting the F's

Clearly the first two F's of Table 10.9 are not significant, for they are less than 1.0. We shall turn to the F for interaction. Entering the Table

Table 10.9

Summary of the Analysis of Variance for the Sex-IQ Experiment

Source of Variation	Sum of Squares	df	Mean Square	F
Between sex	2.01	1	2.01	0.73
Between IQ	0.82	1	0.82	0.30
Interaction S × I	1411.35	1	1411.35	513.22
Within groups	154.00	56	2.75	
Total	1568.18	59		

of F with one and 56 df we find that we must interpolate between 40 and 60 df. Remembering that we set our significance level at 0.01, we interpolate between 7.31 and 7.08. The resulting value of F is 7.13. Since our F of 513.22 for interaction exceeds this value (and it certainly does), we may conclude that our F has a P less than 0.01. As a result of these findings we fail to reject our first two null hypotheses, but reject the third. Hence, we may conclude that neither variation of sex nor variation of intelligence (in general) influences problem-solving ability. But this statement must be tempered, since there is a significant interaction between these two variables. We may interpret this finding as follows: whether or not sex influences problem-solving ability depends on the level of intelligence of the subjects. Or equally, whether or not intelligence influences problem-solving ability depends on the sex of the subject. This relationship is shown in Figure 10.15, where the means of the four conditions are plotted.

Recalling that the fewer the number of questions asked the better the performance, we note in Figure 10.15 that the high IQ males did considerably better than the high IQ females, but the low IQ females did considerably better than the low IQ males. The above interpretation of the interaction should be helped by the diagram in Figure 10.15. If someone asked "Who is better, males or females?" there could be given no simple answer based on our tabulation, for it is not true that either males, in general, or females, in general, are significantly superior. The best that could be said is it depends on the intelligence of the person. If the consideration is restricted to people of high intelligence, then "males" is the answer; or, if one is only concerned with individuals of low IQ, "females" is the answer. But if one won't so restrict himself, all you can say is: high IQ males are superior to high IQ females, but low IQ females are superior to low IQ males.

This completes our examples of the statistical analysis of factorial designs. We have discussed factorial designs generally, but have only illustrated the analysis for the 2 × 2 case. If you are interested in obtaining general principles for the analysis of any factorial design you should consult one of the references previously given or plan on taking a

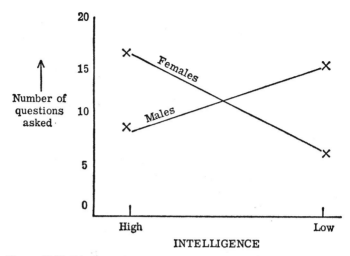

Figure 10.15. Diagramatic representation of a hypothetical interaction between sex and intelligence.

more advanced course. It is not likely, however, that you will get beyond the 2 × 2 design in your elementary work.

The Choice of a Correct Error Term

One of the most important problems of statistical analysis is the choice of the correct error term. With reference to the *F* test the problem is one of choosing the correct denominator. The error term that we have used is the within-groups mean square. While the within-groups mean square is usually the correct error term, we should be aware that sometimes it is not. To understand this, let us note that there are three types of factorial designs that you are likely to encounter in your work. We shall refer to the first as the case of fixed variables, the second as the case of random variables, and the third as the case of mixed variables. ("Variables" in each case refers to the independent variables of the factorial design.) Let us take these three cases in turn and illustrate them by means of a 2 × 2 design.

The Case of Fixed Variables

The 2 × 2 design indicates that we have two independent variables, each varied in two ways. Now, if we have some particular reason to select the two values of the two variables, it can be said that we are dealing with fixed variables. This is so because we have not arrived at the particular values in a random manner. In other words, we have chosen the two values of each independent variable in a premeditated way. We are interested in method A of teaching (a *specific* method) versus method B, for example. Or we choose to study 10 hours of training versus 20 hours. Similarly, we decide to give our rats 50 versus 100 trials, selecting these particular values for a particular reason. When we select our values of the independent variables for some specific reason, and do not select them at random, we have the case of fixed variables. *For this case the within-groups mean square is the correct error term for all F tests being run.* If we refer to our two independent variables as K and L, and the interaction as K × L, we have the following between-groups mean squares to test: that between the two conditions of K, that between the two conditions of L, and that for K × L. For the case of fixed variables, each of the between-groups mean squares should be divided by the within-groups mean square. Since this is the case most frequently encountered in psychological research, we have chosen to illustrate this case in our previous examples in the present chapter. For if you review our examples you will note that we intended to select the values of our independent variables for a particular reason, rather than at random. This point will become clearer as we turn to the second case.

The Case of Random Variables

If the values of the two independent variables have been selected at random, you are using the case of random variables. For example, if our two variables are number of trials and IQ of subjects, we would consider all possible reasonable numbers of trials and all possible reasonable IQ's. Our two particular values of each independent variable would then be selected strictly at random. For instance, we might consider as reasonable possible values of the first independent variable—numbers of trials— those from 6-300. We would then place these 295 numbers in a hat and draw two from them. The resulting numbers would be the values that we would assign to our independent variable. The same process would be followed with regard to the IQ variable. In the case of studying various characteristics of subjects such as IQ, however, a more suitable procedure could be followed. If a random sample of subjects has been drawn, then merely grouping subjects into classes would satisfy our requirement. That is, we might divide all subjects into two groups, using a certain IQ score as the dividing line. This would constitute random values of this independent variable. The reason that this is so is that we have specified that our subjects were selected at random from a given population. Hence, in randomly selecting subjects, we also randomly select values of their characteristics, such as IQ.

The procedure for testing the between-groups means squares for the case where both *independent variables are random variables is as follows: test the interaction mean square by dividing by the within-groups mean square. Then test the other mean squares by dividing by the interaction mean square.* That is, test the K × L mean square by dividing it by the mean square within groups. Then test the mean square between the two conditions of K by dividing it by the K × L mean square, and also test the mean square between L by dividing it by the K × L mean square. We might remark that designs in which both variables are random are relatively rare in psychological research.

The Case of Mixed Variables

This is a less uncommon case than that where both variables are random, but still does not occur as frequently as the case of fixed variables. The case of mixed variables occurs when one independent variable is fixed and the other is random. The procedure for testing the three mean squares for this case is as follows: *divide the within-groups mean square into the interaction mean square; divide the interaction mean square into the mean square for the fixed independent variable; and divide the within mean square into the mean square for the random independent variable.*

We have covered the choice of the correct error term for three types of factorial designs, although there are a number of variations that can occur. We have not discussed these variations since they occur even less frequently than the above situations.[7]

The importance of the above discussion may not be immediately apparent to you beyond knowing that there is a "right" way and a "wrong" way to analyze the various cases. You will, however, recognize the importance of the type of factorial design that you select and the choice of the correct error term when we come to the topic of generalization in Chapter 13.

The Importance of Interactions

Our goal in psychology is to arrive at statements about the behavior of organisms which can be used for such purposes as explaining, predicting, and controlling behavior. And for accomplishing these purposes we would like our statements to be as simple as possible. However, we must recognize that behavior is anything but simple. Therefore it should be very

---

[7] For a more thorough consideration of this topic you are referred to Anderson and Bancroft (1952, Chapter 23), or if you prefer a psychological text, to Lindquist (1953).

surprising to us if our *statements* about behavior are simple. It would seem more reasonable to expect that complex statements must be made about complex events. Those who talk about behavior in simple terms are likely to be wrong. This is illustrated by "common sense" discussions of behavior. People often say such things as "he is smart, he will do well in college," or "she is pretty, she will have no trouble finding a husband." However, such matters are not that uncomplicated—there are variables other than "smartness" that influence how well a person does in college, and there are variables other than beauty that influence a girl's marriageability. Furthermore, such variables do not always act on all people in the same manner. Rather, they *interact* in such a way that people with certain characteristics behave one way, but people with the same characteristics in addition to other characteristics behave another way.

Consider two variables that might influence the likelihood of a girl's getting married: beauty and intelligence. Then consider two values of each of these variables: beautiful and not beautiful, high intelligence and low intelligence. We could assign a sample of girls to four groups on which we could collect data: beautiful girls with high intelligence, beautiful girls with low intelligence, not-beautiful girls with high intelligence, and not-beautiful girls with low intelligence. Suppose we collected data on the frequency with which girls in these four groups married, and found that: beautiful girls with low intelligence get married most frequently, not-beautiful girls with high intelligence get married next most frequently, beautiful girls with high intelligence get married with the third greatest frequency, and low-intelligence girls who are not beautiful get married the least frequently (see Figure 10.16).

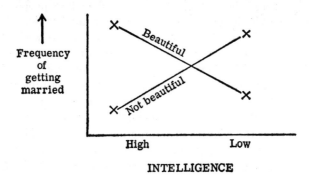

Figure 10.16. A possible interaction between beauty and intelligence.

Now, if these findings were actually obtained, then the simple statement, "she is pretty, she will have no trouble finding a husband" is inaccurate. Beauty is not the whole story—intelligence is also important. We cannot say that beautiful girls are more likely to get husbands any more than we can say that unintelligent girls are more likely to get husbands. The only accurate statement is that beauty and intelligence interact—beautiful girls with low intelligence are more likely to get married than unbeautiful girls with low intelligence; but unbeautiful girls with high intelligence are more likely to get married than beautiful girls with high intelligence. Incidentally, these possible findings appear to be quite reasonable, at least to the author. For clearly, boys like girls to be beautiful, but we often hear (perhaps erroneously) that "boys don't like girls to be too intelligent." Hence the superiority of low intelligence-beautiful girls. High intelligence girls who are not beautiful might be next, for they are smart enough to compensate for their lack of beauty, in addition to the possibility that high intelligence girls who are beautiful might just "scare" boys away with their "dazzling" combination. In any event, the answer to these questions must await suitable research.

We have just barely begun to make completely accurate statements when we talk about interactions between two variables. For it is highly likely that interactions of a much higher order occur, that is, interactions among three, four, or any number of variables. To illustrate, not only might beauty and intelligence interact, but in addition such variables as desire to get married (too high a desire might get the boys "worried" too soon), economic status of the parents, social graces, and so on. Hence, for a really adequate understanding of behavior, we need to determine the large number of interactions that undoubtedly occur. In the final analysis, if such ever occurs in psychology, we will probably arrive at two general kinds of statements about behavior: those statements that tell us how everybody behaves (those ways in which people are similar), with no real exceptions; and those statements that tell us how people differ. The latter will probably involve statements about interactions. For people with certain characteristics act differently than people with other characteristics in the presence of the same stimuli. And statements that describe the varying behavior of people will probably rest on accurate determination of interactions. If such a complete determination of interactions ever comes about, we will be able to understand the behavior of what humanists call the "unique" personality.

Now let us refer the concept of interaction back to a topic discussed in Chapter 2. We discussed ways in which we become aware of a problem, one of them being as a result of contradictory findings in a series of experiments. For example, we considered two experiments, each using the same design and performed on the same problem, but with contradictory results. Why? One reason might be that a certain variable is not controlled in either experiment. Hence, it might assume one value in the first experiment and a second value in the second experiment. And if such an extraneous variable interacts with the independent variable(s), then the discrepant results become understandable. A new experiment could then be conducted in which that extraneous variable becomes an independent variable. As it is purposively manipulated along with the original independent variable, the nature of the interaction can be determined. In this way not only will the apparently contradictory results be understood, but a new advance in knowledge will be made.

This situation need not be limited to the case where the extraneous variable is uncontrolled. For instance, the first experimenter may hold the extraneous variable constant at a certain value while the second experimenter may also hold it constant, but at a different value. And the same result would obtain as when the variable went uncontrolled—contradictory findings in the two experiments. Let us say we are interested in whether a democratic or an authoritarian approach to teaching is better. So we conduct an experiment. We find that students learn more with a democratic approach. Experimenter Jones, however, repeats our experiment and finds that his subjects learned better when the authoritarian approach was used. Why? Perhaps our subjects are more intelligent than Jones'. But in both experiments care was taken to control this variable—intelligence was held constant for both groups. But it may have been controlled at different values. Hence we repeat the experiment, but obtain a somewhat wider sample of subjects as far as IQ's are concerned. In effect, we attempt to repeat both his and our experiments, but systematically varying intelligence in two ways. The design, a 2×2 factorial, would look like that in Figure 10.17a.

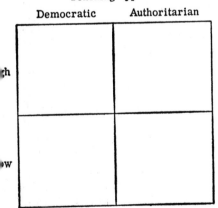

Teaching approach

Democratic   Authoritarian

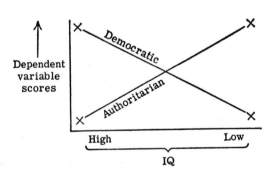

Dependent variable scores

High   Low

IQ

Figure 10.17a. A design to systematically investigate the effect of an extraneous variable.

Figure 10.17b. Possible interaction between teaching approach and IQ.

Now let us say that the results in the new (factorial) experiment come out as they did in the two original ones: the democratic approach leads to higher learning for the high intelligence subjects (our original experiment), but the reverse is true for the low intelligence subjects (Jones'

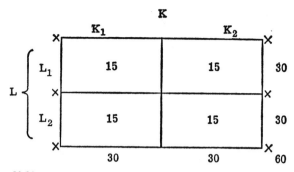

Figure 10.18. A 2 × 2 design showing the numbers of subjects for cells, conditions, and the total number in the experiment.

experiment). We have determined that an interaction exists between type of approach to teaching and intelligence (see Figure 10.17b), and what originally looked like a contradiction is resolved.

Undoubtedly these considerations hold for a wide variety of experimental findings, for the contradictions in the psychological literature are legion. By shrewd application of factorial designs to such problems their resolution should be accomplished.

Value of the Factorial Design

Not long ago the standard design in psychological research was the two-groups design. For many years, however, statisticians and researchers in such fields as agriculture and genetics, had been developing other kinds of designs. One of these was the factorial design, which, incidentally, grew with the development of analysis of variance. Slowly, psychologists started trying out these designs in their own area. Some of them were found to be inappropriate, but the factorial design is one that has enjoyed success, and the extent of its success is widening. It is particularly applicable to psychological problems, although some psychologists still hold that the two-groups design should be the standard for our kinds of research. Our position has been that each type of design that we have considered is appropriate for particular situations. We cannot say that one design should *always* be used, and that the others are useless. However, we would make the general statement that where it is feasible, the factorial design is superior to the other designs that we discussed. Professor Ronald Fisher has made some interesting comments on this point.

In expositions of the scientific use of experimentation it is frequent to find an excessive stress laid on the importance of varying the essential conditions *only one at a time*. The experimenter interested in the causes which contribute to a certain effect is supposed, by a process of abstraction, to isolate these causes into a number of elementary ingredients, or factors, and it is often supposed, at least for purposes of exposition, that to establish controlled conditions in which all of these factors except one can be held constant, and then to study the effects of this single factor, is the essentially scientific approach to an experimental investigation. This ideal doctrine seems to be more nearly related to expositions of elementary physical theory than to laboratory practice in any branch of research. In experiments merely designed to illustrate or demonstrate simple laws, connecting cause and effect, the relationships of which with the laws relating to other causes are already known, it provides a means by which the student may apprehend the relationship, with which he is to familiarize himself, in as simple a manner as possible. By contrast, in the state of knowledge or ignorance in which genuine research, intended to advance knowledge, has to be carried on, this simple formula is not very helpful. We are usually ignorant which, out of innumerable possible factors, may prove ultimately to be the most important, though we may have strong presuppositions that some few of them are particularly worthy of study. We have usually no knowledge that any one factor will exert its effects independently of all others that can be varied, or that its effects are particularly simply related to variations in these other factors. On the contrary, when factors are chosen for investigation, it is not because we anticipate that the laws of nature can be expressed with any particular simplicity in terms of these variables, but because they are variables which can be controlled or measured with comparative ease. *If the investigator, in these circumstances, confines his attention to any single factor, we may infer either that he is the unfortunate victim of a doctrinaire theory as to how experimentation should proceed, or that the time, material or equipment at his disposal is too limited to allow him to give attention to more than one narrow aspect of his problem.* The modifications possible to any complicated apparatus, machine or industrial process must always be considered as potentially interacting with one another, and must be judged by the probable effects of such interactions. If they have to be tested one at a time this is not because to do so is an ideal scientific procedure, but because to test them simultaneously would sometimes be too troublesome, or too costly. In many instances, the belief that this is so has little foundation. Indeed, in a wide class of cases an experimental investigation, at the same time as it is made more comprehensive, may also be made more efficient if by more efficient we mean that more knowledge and a higher degree of precision are obtainable by the same number of observations. (Fisher, 1953, pp. 91-92, Italics ours.)

Let us look into this matter more thoroughly. First we may note that the amount of information obtained from a factorial design is considerably greater than that obtained from the other designs mentioned, when we consider the number of subjects used. For example, let us say that we have two problems: (1) does variation of independent variable K affect a given dependent variable; and (2) does variation of independent variable L affect that same dependent variable. If we investigated these two problems by the use of a two-groups design, we would obtain two values for each variable. With sixty subjects available for each experiment, the design for the first problem would be:

Experiment #1

Group K_1
30 subjects

Group K_2
30 subjects

And similarly for the second problem:

Experiment #2

Group L₁
30 subjects

Group L₂
30 subjects

With a total of 120 subjects we are able to evaluate the effect of the two independent variables. We would not be able to tell if there is an interaction between A and B if we looked at these as two separate experiments.

But what if we used a factorial design to answer our two problems? Assume that we still want 30 subjects for each condition. In this case the factorial would be as in Figure 10.18. We would have four groups with

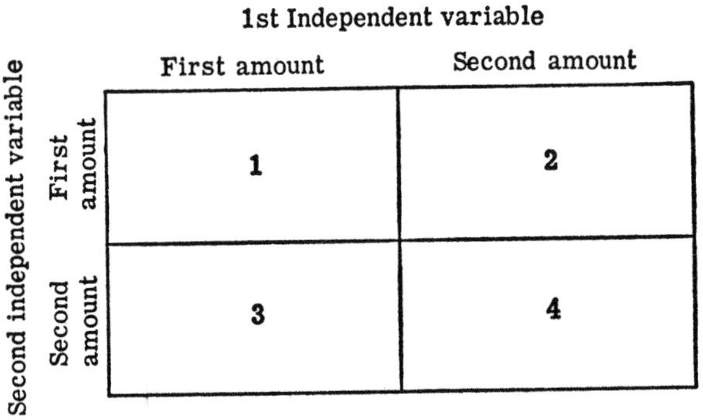

1st Independent variable

	First amount	Second amount
First amount	1	2
Second amount	3	4

(Second independent variable)

Figure 10.19. A general representation of a 2 × 2 factorial design.

15 subjects per group. But for comparing the two conditions of K we would have 30 subjects for condition K₁ and 30 subjects for K₂. This is just what we had for experiment #1. And the same for the second experiment: we have 30 subjects for each condition of L. Here we have accomplished everything with the 2 × 2 factorial design that we would have accomplished with the two separate experiments with two groups. But with those two experiments we required 120 subjects in order to have 30 available for each condition; however, with the factorial design we need only 60 subjects, to have the same number of subjects for each condition. The factorial design is much more efficient because we use our subjects simultaneously for testing both independent variables.[8] In addition, we can evaluate the interaction between K and L—something that we did not do for the two two-groups experiments. While we may look at the information about the interaction as pure "gravy," we should note that

[8] We are assuming that the two two-groups experiments are analyzed independently. Otherwise, the error terms might be pooled, which would result in a larger number of degrees of freedom than for the factorial design.

some hypotheses may be constructed specifically to test for interactions. Thus, it may be that the experimenter is primarily interested in the interaction, in which case the other information may be regarded as "gravy." But whatever the case, it is obvious that the factorial design yields considerably more information than separate two-groups designs, and at considerably less cost to the experimenter. Still other advantages of the factorial design could be elaborated, but they might well come in your more advanced courses.

Summary of an Analysis of Variance and the Computation of an F test for a 2 X 2 Factorial Design

Assume that the following dependent variable scores have been obtained for the four groups in a 2 × 2 factorial design.

Condition A

	A_1	A_2
B_1	2	3
	3	4
	4	5
	4	7
	5	9
	6	10
	7	13
B_2	5	4
	6	6
	7	7
	8	9
	8	10
	8	11
	8	14

Condition B:

1. The first step is to compute ΣX, ΣX^2, and n for each condition. The values have been computed for our example:

Condition A

	A_1	A_2
B_1	$\Sigma X = 31$	$\Sigma X = 51$
	$\Sigma X^2 = 155$	$\Sigma X^2 = 449$
	$n = 7$	$n = 7$
B_2	$\Sigma X = 50$	$\Sigma X = 61$
	$\Sigma X^2 = 366$	$\Sigma X^2 = 599$
	$n = 7$	$n = 7$

Condition B

2. Using equation 10.1, we next compute the total SS:

$$\text{Total } SS = (\Sigma X_1^2 + \Sigma X_2^2 + \Sigma X_3^2 + \Sigma X_4^2) - \frac{(\Sigma X_1 + \Sigma X_2 + \Sigma X_3 + \Sigma X_4)^2}{N}$$

$$= (155 + 449 + 366 + 599) - (31 + 51 + 50 + 61)^2/28$$

$$= 238.68$$

3. The over-all between SS is computed by substituting in equation 10.2:

Between SS

$$= \frac{(\Sigma X_1)^2}{n_1} + \frac{(\Sigma X_2)^2}{n_2} + \frac{(\Sigma X_3)^2}{n_3} + \frac{(\Sigma X_4)^2}{n_4}$$
$$- \frac{(\Sigma X_1 + \Sigma X_2 + \Sigma X_3 + \Sigma X_4)^2}{N}$$

$$= (31)^2/7 + (51)^2/7 + (50)^2/7 + (61)^2/7 - (31 + 51 + 50 + 61)^2/28$$

$$= 67.25$$

4. The within SS is determined by substraction (equation 10.3):

$$\text{Total } SS - \text{Over-all between } SS = \text{Within } SS$$
$$238.68 - 67.25 = 171.43$$

5. We now seek to analyze the Over-all between SS into its components, viz., the between A SS, the between B SS and the $A \times B$ SS. The Between A SS may be computed with the use of equation 10.4.

Between A SS

$$= \frac{(\Sigma X_1 + \Sigma X_3)^2}{n_1 + n_3} + \frac{(\Sigma X_2 + \Sigma X_4)^2}{n_2 + n_3} - \frac{(\Sigma X_1 + \Sigma X_2 + \Sigma X_3 + \Sigma X_4)^2}{N}$$

$$= \frac{(31 + 50)^2}{7 + 7} + \frac{(51 + 61)^2}{7 + 7} - \frac{(31 + 51 + 50 + 61)^2}{28} = 34.32$$

The Between B SS may be computed with the use of equation 10.5.

Between B SS

$$= \frac{(\Sigma X_1 + \Sigma X_2)^2}{n_1 + n_2} + \frac{(\Sigma X_3 + \Sigma X_4)^2}{n_3 + n_4} - \frac{(\Sigma X_1 + \Sigma X_2 + \Sigma X_3 + \Sigma X_4)^2}{N}$$

Between B SS

$$= \frac{(31 + 51)^2}{7 + 7} + \frac{(50 + 61)^2}{7 + 7} - \frac{(31 + 51 + 50 + 61)^2}{28} = 30.04$$

The sum of squares for the interaction component $(A \times B)$ may be computed by subtraction:

$$A \times B \ SS = \text{Over-all Between } SS - \text{Between } A \ SS - \text{Between } B \ SS$$
$$67.25 - 34.32 - 30.04 = 2.89$$

6. Compute the several degrees of freedom. In particular, determine df for the total source of variance (equation 10.7), for the overall between source (equation 10.8) and the within source (equation 10.9). Following this, allocate the over-all between degrees of freedom to the components of it; namely that between A, that between B, and that for $A \times B$.

$$\text{Total } df = N - 1$$
$$= 28 - 1 = 27$$
$$\text{Over-all Between } df = r - 1$$
$$= 4 - 1 = 3$$
$$\text{Within } df = N - r$$
$$= 28 - 4 = 24$$

The components of the over-all between df are:

$$\text{Between } df = r - 1$$
$$\text{Between } A = 2 - 1 = 1$$
$$\text{Between } B = 2 - 1 = 1$$
$$A \times B \ df = (\text{number of } df \text{ for between } A) \times (\text{number of } df \text{ for between } B)$$
$$= 1 \times 1 = 1$$

7. Compute the various mean squares. This is accomplished by dividing the several sums of squares by the corresponding degrees of freedom. For our example these operations, as well as the results of the preceding ones are summarized:

Summary of the Analysis of Variance for Our Example

Source of Variation	Sum of Squares	df	Mean Square	F
Between A	34.32	1	34.32	4.81
Between B	30.04	1	30.04	4.21
$A \times B$	2.89	1	2.89	.40
Within groups	171.43	24	7.14	
Total	238.68	27		

8. Compute an F for each "Between" source of variation. In a 2×2 factorial design there are three F tests to run. The F is computed by dividing a given mean square by the within groups mean square (assuming the case of fixed variables). These F's have been computed and entered in the above table.

9. Enter Table 9.10 to determine the probability associated with each F. To do this find the column for the number of degrees of freedom associated with the numerator and the row for the number of degrees of freedom associated with the denominator. In our example they are 1 and 24 respectively. The F of 4.81 for Between A would thus be significant

beyond the 5 per cent level, and accordingly we would reject the null hypothesis for this condition. The F Between B (4.21) and that for the interaction (0.40), however, are not significant at the 5 per cent level; hence we would fail to reject the null hypotheses for these two sources of variation.

Problems

1. An experimenter wants to evaluate the effect of a new drug on "curing" psychotic tendencies. He investigates two independent variables, the amount of the drug administered and the type of psychotic condition. He decides to vary the amount of drug administered in two ways, none and 2 cc. The type of psychotic condition is also varied in two ways, schizophrenic and manic-depressive. Diagram the factorial design that he used.

2. In the above experiment the psychologist used a measure of normality as his dependent variable. This measure varies between zero and 10, where 10 is very normal and zero is very abnormal. Seven subjects were assigned to each cell. The resulting scores for the four groups were as follows. Conduct the appropriate statistical analysis and reach a conclusion about the effect of each variable and the interaction.

Psychotic Condition

Schizophrenics		Manic Depressives	
Received Drug	Did Not Receive Drug	Received Drug	Did Not Receive Drug
6	2	5	1
6	3	6	1
6	3	6	2
7	4	7	3
8	4	8	4
8	5	8	5
9	6	9	6

3. How would the above design be diagrammed if the experimenter had varied the amount of drug in three ways (zero amount, 2 cc, and 4 cc), and the type of psychotic tendency in three ways (schizophrenic, manic-depressive, and paranoid)?

4. How would you diagram the above design if the experimenter had varied the amount of drug in four ways (zero, 2 cc, 4 cc, and 6 cc) and the type of subject in four ways (normal, schizophrenic, manic-depressive, and paranoid)?

5. A cigarette company is interested in the effect of several conditions of smoking on steadiness. They manufacture two brands, Old Zinc and

Counts. Furthermore they make each brand with and without a filter. A psychologist conducts an experiment in which he studies two independent variables. The first is brand, which is varied in two ways (Old Zinc's and Counts), and the second is filter, which is also varied in two ways (with a filter and without a filter). He uses a standard steadiness test as his dependent variable. Diagram the resulting factorial design.

6. In the above experiment the higher the dependent variable score, the greater the steadiness. Assume that the results came out as follows (10 subjects per cell). What conclusions did the experimenter reach?

Old Zinc		Counts	
With Filter	*Without Filter*	*With Filter*	*Without Filter*
7	2	2	7
7	2	3	7
8	3	3	7
8	3	3	8
9	3	3	9
9	4	4	9
10	4	4	10
10	5	5	10
11	5	5	11
11	5	6	11

7. An experiment is conducted to investigate the effect of opium and marijuana on hallucinatory activity. Both of these independent variables are varied in two ways. Seven subjects were assigned to cells and amount of hallucinatory activity was scaled so that a high number indicates considerable hallucination. Assuming that adequate controls have been realized, and that a 5 per cent level of significance was set, what conclusions can be reached?

Smoked Opium		Did Not Smoke Opium	
Smoked Marijuana	*Did Not Smoke Marijuana*	*Smoked Marijuana*	*Did Not Smoke Marijuana*
7	5	6	3
7	5	5	2
7	4	5	2
6	4	4	1
6	3	4	1
5	3	4	0
4	3	3	0

II

The Logical Bases of
Experimental Inferences

We have said that an experimenter should state his hypothesis explicitly. His experiment has as its purpose the gathering of data that are relevant to the hypothesis. These data are summarized in the form of an evidence report. The experimenter confronts the hypothesis with the evidence report. If the two are in agreement, he concludes that the hypothesis is confirmed. Otherwise it is not. Let us analyze the relationship between the hypothesis and the evidence report.

The hypothesis, preferably, is stated in the conditional form; that is, as an if-then relation. To pursue the example of Chapter 3, recall that the hypothesis that industrial work groups in great inner conflict have low production levels was stated as follows: *If* an industrial work group is in great inner conflict, *then* that work group will have a low production level. To test this hypothesis we might form two groups of subjects, one in great conflict but the other quite harmonious. We would then collect data on the production output of the two groups. Assume that the statistical analysis of the data shows that the in-conflict-group has a significantly lower production level than the harmonious group.

Forming the Evidence Report[1]

An evidence report is a summary statement of the results of an empirical investigation—it is a sentence that precisely summarizes what was found. In addition, the evidence report states that the antecedent conditions of the hypothesis were realized. Hence, the evidence report consists of two parts: a statement that the antecedent conditions of the hypothesis held, and that the consequent conditions were found to be either true or false. The general form for stating the evidence report is

[1] The term "evidence report" is taken from Hempel (1945) and is used because of its descriptive nature. Similar terms are "observational sentence," "protocol sentence" and "concept by inspection."

244

thus that of a conjunction. Recalling the general form of the hypothesis as "If *a*, then *b*," *a* denotes the antecedent conditions of the hypothesis and *b* the consequent conditions. Hence, the possible evidence reports would be "*a* and *b*," or "*a* and not *b*," for the cases where the consequent conditions were found to be (probably) true and false respectively. The former is a positive evidence report and the latter is a negative one.

We shall illustrate by continuing the example. Let *a* stand for "an industrial work group is in great inner conflict" and *b* for "that work group will have a lowered production level." In our hypothetical experiment we had an industrial work group that was in great inner conflict, and so we may assert that the antecedent conditions of our hypothesis were realized, that they were present in the experimental situation. Since our finding was that that work group had a lower production level than a control group, we may also assert that the consequent conditions were found to be true. Thus, our evidence report is: "An industrial work group was in great inner conflict *and* that work group had a lowered production level."

At this point, let us examine how we tell whether the consequent conditions of the hypothesis are true or false. In this example the group with inner conflict had a lower production level than did a control group. We needed the control group as a basis of comparison. For without such a basis "lower production level" does not mean anything—it must be lower than something. And so it is for all experiments. The way to tell whether or not consequent conditions are true is by comparing the results for the experimental group with those for some other group, usually a control group. That the hypothesis implicitly assumes the existence of a control or other group may be clarified by stating the hypothesis in the following manner: "If an industrial work group is in great inner conflict, then that work group will have a lower production level *than that of a group that is not in inner conflict.* And the direction of the comparison is determined by the consequent conditions of the hypothesis. In this example, the hypothesis states that the group with inner conflict should have a *lower* production level than a control group. If the statistical analysis indicates that it is significantly lower than the control group, we conclude that the consequent conditions are probably true. If the statistical analysis indicates that the group with the inner conflict has a production level that is significantly higher than the control group, however, or if there is no significant difference between the two levels, we conclude that the consequent conditions are probably false. The evidence report would then be: "An industrial work group was in great inner conflict *and* that work group did not have a lowered production level."

With this format for forming the evidence report before us, we shall now consider the nature of the inferences made from it to the hypothesis. Before we do this, however, it will be necessary to discuss the general topic of inferences and how they are made.

Inductive and Deductive Inferences

Let us say that we have a set of statements that we shall denote by A. These statements contain information on the basis of which we can reach another statement, B. Now, when we proceed from A to B, we make an *inference*. An inference, then, is a process by which we reach a conclusion on the basis of certain preceding statements—it is a process of reasoning whereby we start with A and arrive at B. We then have some belief that B is true on the basis of A. There are two kinds of inferences, inductive and deductive. In both types, our belief in the truth of B is based on the assumption that A is true. The essential difference between the two is that in an inductive inference, we reach the conclusion that B is true with some degree of probability; with a deductive inference, however, we can conclude that B is *necessarily* true if A is true.

The statement "Every morning that I have arisen, I have seen the sun rise" might be A. On the basis of this statement, we may infer the statement B: "The sun will always rise each morning." Now, does B necessarily follow from A? It *does not*, for while you may have always observed the rise of the sun in the past, it does not follow that it will *always* rise in the future. B is not *necessarily* true on the basis of A. Although it may seem unlikely to you now, it is entirely possible that one day, regardless of what you have observed in the past, the sun will not rise. We can only say then, that B has some degree of probability (is probably true) on the basis of the information contained in A. When we make an inference that may be in error, we say that the result of the inference (the conclusion) can only have a certain probability of being true. Thus, the probability that statement B can be (inductively) inferred to be true on the basis of statement A may be high, medium, or low. The fact that we make inferences from one statement to another with a certain degree of probability sometimes leads us to use the term *probability inference* as a synonym for *inductive inference*.

It is possible (at least in principle) to determine the probability of an inference precisely as a specific number, rather than simply to say "high" or "low." Conventionally, the probability of an inductive inference may be expressed by any number from zero to one.[2] Thus, we may say that the probability (P) of the inference from A to B is 0.40, or 0.65, or whatever. Furthermore the closer P is to 1.0, the higher the probability that the inference will result in a true conclusion (again, assuming that A is

[2] To emphasize that 0 to 1 is an arbitrary range we may note that some authors allow P to assume any value between -1 and $+1$.

true). By analogy, the closer P is to 0.0, the lower the probability that the inference will result in a true conclusion—or, if you will, the higher the probability that the inference will result in a false conclusion. At this point we may note that if the probability of an inference is 1.0, the conclusion is necessarily true if the statements on which it is based (i.e., A) are true; and if P is 0.0, the conclusion is necessarily false. As we have previously noted, however, neither of these situations obtain when an inductive inference is made.

We may thus say that if the probability that B follows from A is 0.99, it is rather sure that B is true. The previous example of the inference from "Every morning that I have arisen I have seen the sun rise" to "the sun will always rise each morning" is an example of an inference with a high probability—the probability of this inference is, in fact, extremely close to, but still not quite, 1.0. On the other hand, if the probability of the inference from A to B is 0.03, we know that this probability is extremely low and thus are not likely to accept B as true on the basis of A. For example, the probability of the inference from the statement "a person has red hair" to the conclusion that "that person is very temperamental," would be one with a very low probability.

Summary

In short, then, an inductive inference is made when one passes from one statement to another with a lack of certainty—he infers that one statement probably implies another. And we may express our degree of belief in the truth of the results of such an inference by the use of a number that varies from 0.0 to 1.0. The higher the number, the greater the probability that the inference results in a true conclusion.

We said that a deductive inference is made when the truth of one statement is necessary, based on another one or set of statements, if statement A necessarily implies B. In this case the inference is strict. Consider the following statements as an example of a deductive inference. We might know that "all anxious people bite their nails" and further that "John Jones is anxious." We may, therefore, deductively infer that "John Jones bites his nails." In this example, if the first two statements are true (they are called premises), the final statement (the conclusion) is necessarily true.

The determination of whether or not a given inference is deductive or inductive lies in the realm of logic. In both deductive and inductive (probability) logic certain rules (known as rules of inference or transformation) have been developed that indicate how to proceed from one set of statements to another. If an inference conforms to the rules of deductive logic, it is deductive. If it conforms to the rules of inductive logic, it is inductive. Since this is not a book in logic, we shall not explore the various rules to any great extent. We shall simply indicate the rules that are used in making inferences from evidence reports to various types of hypotheses. In doing this, we shall indicate which are valid inferences, and whether they are inductive or deductive.

Direct vs. Indirect Statements

The statements with which science deals may be divided into two categories, direct or indirect. A direct statement is one that refers to limited phenomena that are immediately observable, that is, phenomena that can all be observed directly with the senses. For example, the statement "that bird is red" is direct. Of course the use of various kinds of auxiliary apparatus (e.g., microscopes, telescopes) to aid the senses may be used in forming direct statements. Hence, the statements "there is a sun spot" or "there is an amoeba" are also direct statements, since a person's sensory apparatus can be extended by various types of equipment. The procedure for testing a direct statement is straightforward: compare the statement with an observation of the phenomenon with which it is concerned. More precisely, compare the statement with the evidence report. If the two are in agreement, the statement may be regarded as true; otherwise it is false. For example, in testing the direct statement "That door is open," we observe the door. If we find that it *is* open, our observation agrees with the direct statement and we conclude that the statement is true. If we observe the door to be closed, we conclude that the direct statement is false.

An indirect statement is one that cannot be directly tested. Such statements usually deal with phenomena that cannot be directly observed (logical constructs, such as atoms, electricity, or habits) or that are so numerous or extended in time that it is impossible to view them all. A universal hypothesis is of this type—"All men are anxious." It is certainly impossible to observe all men (living, dead, and as yet unborn) to see if the statement is true. The universal hypothesis is the type in which scientists are most interested, since it is an attempt to say something about variables for all time, at all places, etc.

Clearly, then, indirect statements cannot be directly tested. In order to test indirect statements it is necessary to reduce them to direct statements. This reduction is accomplished by following the rules for deductive inferences. Consider an indirect statement P. By drawing deductive inferences from P we may arrive at certain logical consequences, which we shall denote p_1, p_2, etc. Now among the statements p_1, p_2, etc., we expect to find at least some that are direct in nature. Such statements may be tested by comparing them with appropriate evidence reports.

Now, if these directly testable consequences of our indirect statement are found to be true, we may make a further (inductive) inference that the indirect statement itself is probably true. That is, while we cannot directly test an indirect statement, we can derive deductive inferences from such a statement and directly test them. If such directly testable statements turn out to be true, we may inductively infer that the indirect statement is probably true (see note 1 in the Appendix). But if its consequences turn out to be false, we must infer that the indirect statement is also false. In short, indirect statements that have true consequences are themselves probably true, but indirect statements that have false consequences are themselves false. This procedure is represented in Figure 11.1, although it will be necessary to analyze it more thoroughly later.

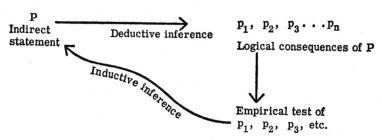

Figure 11.1. Representation of the procedure for testing indirect statements.

To illustrate this procedure let us consider the universal hypothesis "All men are anxious." Assume that we know that "John Jones is a man," and "Harry Smith is a man." From these statements (premises) we can state (by deductive inference) that "John Jones is anxious" and "Harry Smith is anxious." Since the universal hypothesis is an indirect statement, it cannot be directly tested. However, the deductive inferences derived from this indirect statement *are* directly testable. We only need to determine the truth or falsity of these direct statements. If we perform suitable empirical operations and thereby conclude that the several direct statements are true, we may now conclude, by way of an *inductive inference*, that the indirect statement is confirmed.

The class of variables with which the indirect statement deals is of infinite number. For this reason it is impossible to test all the logically possible consequences of that indirect statement (e.g., we cannot test the hypothesis for all men). Further, it is impossible to make a deductive inference from the direct statements back to the indirect statement—rather, we must be satisfied with an inductive inference. And we know that an inductive inference is liable to error—its probability is less than

1.0. As long as we seek to test indirect statements, we must be satisfied with a probability estimate of their truth. We will never know absolutely that they are true.

Confirmation vs. Verification

Our goal is to determine whether a given universal statement is true or false. To accomplish this goal we reason thusly: *If* the hypothesis is true, *then* the direct statements that are the result of deductive inferences are also true. Now, if we find that the evidence reports are in accord with the logical consequences (the direct statements), we conclude that the logical consequences are true. And if the logical consequences are true, we inductively infer that the hypothesis itself is probably true.

Note that we have been cautious and limited in our statements about concluding that a universal hypothesis is false. Under certain circumstances it is possible to conclude that a universal hypothesis is strictly false (not merely improbable or probably false) on the assumption that the evidence report is reliable. More generally (i.e., with regard to any type of hypothesis), it can be shown that under certain circumstances it is possible strictly to determine that a hypothesis is true or false, rather than probable or improbable—but always on the assumption that the evidence report is true. We will here distinguish between the processes of *verification* and *confirmation*. By verification we mean a process of attempting to determine that a hypothesis is strictly true or strictly false; confirmation an attempt to determine whether a hypothesis is probable or improbable. This ties in with the distinction between inductive and deductive inferences. Under certain conditions it is possible to make a deductive inference from the consequence of a hypothesis (which has been determined to be true or false) back to that hypothesis. Thus, where it is possible to make such a deductive inference, we are able to engage in the process of verification. Where we must be restricted to inductive inferences, the process of confirmation is used. To enlarge on this matter, let us now turn to a consideration of the ways in which the various types of hypotheses are tested (see note 2 in the Appendix).

Inferences from the Evidence Report to the Hypothesis

Universal Hypotheses. Recall that the universal hypothesis specifies that all things referred to in the hypothesis have a certain characteristic. For such hypotheses we have indicated a preference for the "If *a*, then *b*" form. With such a form it is understood that we are referring to all *a* and all *b*. For example, if *a* stands for "rats are reinforced at the end of their maze run" and *b* for "those rats will learn to run that maze with no errors," it is understood that we are talking about *all* rats and *all* mazes on which those rats might be trained. The general procedure for testing this hypothesis may be represented as follows (see note 3 in the Appendix):

Hypothesis: If *a* then *b*
Evidence Report: *a* and *b*
↓ (Inductive Inference)
Conclusion: "If *a* then *b*" is probably true.

To illustrate this procedure by means of our example we might form two groups of rats; Group E is reinforced at the end of each maze run, but Group C is not. Let us say that at the end of a certain number of trials Group E is able to run the maze with no errors, but Group C is still making errors. A *t* test indicates that Group E's performance is significantly superior to that of Group C. We are thus able to assert that the antecedent conditions of the hypothesis were realized and that the data were in accord with the consequent conditions. The evidence report is positive. The inferences involved in the test of this hypothesis may be illustrated as follows:

Universal Hypothesis: If rats are reinforced at the end of their maze runs, then those rats will learn to run that maze with no errors.

Positive Evidence Report: A (specific) group of rats was reinforced at the end of their maze runs and those rats learned to run the maze with no errors.

Conclusion: The hypothesis is probably true.

We have attempted to show some specific steps in testing a hypothesis, to give you some insight into the various inferences that must be made for this purpose. In your actual work, however, you need not specify each step, for that would become cumbersome. Rather, you should simply rely on the brief rules that we present for testing each type of hypothesis. To summarize the rule for testing a universal hypothesis for the case of a positive evidence report, we shall merely say that when the evidence report agrees with the hypothesis, that hypothesis shall be said to be confirmed.

To understand the test of the universal hypothesis when the evidence report is negative, we refer to the distinction between confirmation and verification. Clearly, the case of confronting the universal hypothesis with a positive evidence report is an example of confirmation. When the evidence report is negative, however, we are able to apply the procedure of verification. This is possible because the rules of deductive logic tell us that a deductive inference may be made from a negative evidence report to a universal hypothesis. The procedure for this situation may be illustrated as follows:

> Universal Hypothesis: If a then b
> Evidence Report: a and not b
> $\downarrow$ (Deductive Inference)
> Conclusion: "If a then b" is false

For example:

Universal Hypothesis:	If rats are reinforced at the end of their maze runs, then those rats will learn to run that maze with no errors.
Negative Evidence Report:	A group of rats was reinforced at the end of their maze runs and those rats did not learn to run that maze without any errors.
Conclusion:	The hypothesis is false (see note 4 in the Appendix).

We may thus see that it is possible to determine that a universal hypothesis is (strictly) false (through verification) if the evidence report is negative. But if the evidence report is positive, we cannot determine that the hypothesis is (strictly) true; rather, we can only say that it is probable (through confirmation). This characteristic of being able to verify a hypothesis in only one direction (for example being able to determine that it is strictly false, but not being able to determine that it is strictly true) has been called by Reichenbach (1949) *unilateral verifiability*. And we shall see that it is characteristic of all universal and existential hypotheses.

Existential Hypotheses. This type of hypothesis says that there is at least one thing that has a certain characteristic. Our example, stated as a positive existential hypothesis would be: "There is a (at least one) rat that, if it is reinforced at the end of its maze run, then it will learn to run that maze with no errors." If this type of hypothesis is strictly true, then it can be verified by observing a series of appropriate events until we come upon a positive instance; and a single positive case is sufficient to determine that the hypothesis is true. On the other hand, if the class of variables specified in the hypothesis is infinite in size, or at least indefinitely large, we shall never be able to determine that the hypothesis is strictly false. We can only observe a finite number of events. If, in that finite number of events, we do not observe a positive instance of our hypothesis, we may well expect this state of affairs to continue for future observations. But we will never be sure that a negative instance of the hypothesis will not eventually occur. Hence, we can appreciate that this type of hypothesis is also unilaterally verifiable—it is possible to determine that it is strictly true but not possible to determine that it is strictly false. Let us illustrate the inferences involved in testing the existential hypothesis: For the case of a positive evidence report:

Existential Hypothesis: There is an *a* such that if *a* then *b*
Positive Evidence Report: *a* and *b*
           ↓   (Deductive Inference)
Conclusion: Therefore, the hypothesis is (strictly) true.

For the case of a negative evidence report:

Existential Hypothesis: There is an *a* such that if *a* then *b*
Negative Evidence Report: *a* and not *b*
           ↓   (Inductive Inference)
Conclusion: Therefore, the hypothesis is not confirmed.

To illustrate by means of our previous example:

Existential Hypothesis:	There is a rat that, if it is reinforced at the end of its maze runs, then it will learn to run that maze with no errors.
Positive Evidence Report:	A group of rats was reinforced at the end of their maze runs and at least one of those rats learned to run that maze with no errors.
Conclusion:	The hypothesis is (strictly) true.
Existential Hypothesis:	There is a rat that, if it is reinforced at the end of its maze runs, then it will learn to run that maze with no errors.
Negative Evidence Report:	A group of rats was reinforced at the end of their maze runs and none of those rats learned to run that maze with no errors.
Conclusion:	The hypothesis is not confirmed.

Limited Hypotheses. Limited hypotheses (offered as direct statements) are always verifiable because it is possible to make a deductive inference from the evidence report (whether positive or negative) to the hypothesis. Thus, a limited hypothesis, such as "if rat number 3 is reinforced at the end of its maze runs, then that rat will learn to run that maze with no errors" is completely (bilaterally) verifiable. For instance, if we find that rat number 3 does learn to run the maze with no errors, we may conclude that the hypothesis is true. But if he does not, we may conclude that the hypothesis is false.

Irrelevant Evidence Reports. One final matter needs to be emphasized: the importance of satisfying the antecedent conditions of the hypothesis in the experimental situation. If this requirement is not satisfied, no inference can be made from the evidence report to the hypothesis. Instead, we may only say that the evidence report is irrelevant to the hypothesis and thus does not constitute a test of the hypothesis. For example, if the experimental group of rats in our hypothetical experiment were not reinforced at the end of their maze runs, then whatever the results with regard to the number of errors they make, we cannot say that a test of the hypothesis has occurred.

The Reason for the "If . . . then. . . ." Form

In Chapter 3, we said that universal hypotheses can be made to assume the "If . . . then . . . ." form. After reading this chapter, the reason for our desire to state them in this form, or at least to know that we could state them in this form if we wanted to, should be apparent. To be more explicit, however, let us note that various inferences need to be made from the evidence report to the hypothesis. These inferences can only be made validly if the statements (hypotheses and evidence reports) assume certain forms. For the logical rules that we have borrowed are applicable only to certain forms of statements. By using the conditional form for stating hypotheses, and the conjunctional form for stating evidence reports, we are able to satisfy these rules.

Hence the inferences that we discuss in this chapter are valid, because they conform to the rules of logic. Let us emphasize, however, that strictly speaking, it is not necessary actually to state your hypotheses in the if-then form. All that is required is that, whatever the form in which you state your hypothesis, it is *possible* to restate it in the if-then form. If this is not possible, then it is not possible to make valid inferences from the evidence report to your hypothesis. Similar considerations hold true for the evidence report—in order to perform a valid inference from it to the hypothesis, that evidence report must assume the conjunctional form. If you do not actually state your evidence report in the conjunctional form, however, you need not be concerned *as long as it would be possible to restate it in that form.*

Appendix to Chapter 11

Note #1 (from page 249). It is necessary to point out an error that we are perpetuating in this chapter. It concerns the inductive inference that is made from the evidence report to the hypothesis. Let us consider the universal hypothesis. For this case we said that an inductive inference may be made from a positive evidence report to the hypothesis, an inference that results in the conclusion that the hypothesis is confirmed. But this is not quite right. Rather, it is a procedure that is universally used. To understand the error, let us note that the valid inferences of inductive logic are specified by the rules of probability logic (cf. Reichenbach, 1949). It is not possible to find a rule of this sort in probability.

To make the problem more specific, let us say that we seek to test the hypothesis that "If a then b." Further, assume a positive evidence report was obtained: a and b. Now, we make an inductive inference from that evidence report to the hypothesis, and conclude that the hypothesis is confirmed. However, our hypothesis is not the only hypothesis that will also predict the consequent condition b. Rather, there is an unspecifiable number of additional hypotheses which will also have b as a consequent condition—if a' then b, if a'', then b, if a''', then b, and so on. When we actually find that b is true, then, we do not know which of the numerous hypotheses that imply b is confirmed. While it is customary to assume that it is our hypothesis, it may just as well be one of the other possible hypotheses that is confirmed. About the best we can say at this point is that any hypothesis that implies b may be confirmed, providing that its antecedent conditions were present in the experimental situation. And since there are numerous unspecified antecedent conditions present in the experimental situation, we cannot say that only a was present, for a', a'', a''', etc. may also have been present. Unfortunately we cannot here prolong our discussion of this problem. Let us simply note that there are, however, certain inferences in probability logic that would, in principle, satisfy our needs. However, much additional work needs to be accomplished before we can make these more complicated inferences in our everyday work. For more information on the nature of the difficulty and a proposed solution you might see some earlier work by the present author (McGuigan, 1956).

Note #2 (from page 250). At this point, an apparent contradiction in what we have said might occur to you. We previously stated that it is impossible strictly to determine that a hypothesis is true or false. Rather, we must always settle for a probability statement—that the hypothesis has some degree of probability. Here, however, we have said that we can determine that a hypothesis is false—not merely improbable. How may we reconcile these two statements?

The answer may be stated thus: The deductive inference from the evidence report to the universal hypothesis in effect states that *if* it is true that the evidence report is negative, then the universal hypothesis is necessarily false. But the evidence report itself rests on a probability statement. For we have formed it by failing to reject the null hypothesis. Hence, while we have every reason to believe that the evidence report *really is* negative, we may actually be in error. For this reason we may erroneously conclude that the universal hypothesis is false, even though we are making a deductive inference. Remember that a deductive inference does not lead to an absolutedly irrefutable statement about knowledge. Rather, it allows us to say that *if* the premises are true, then it is necessarily the case that the conclusion is true. But the determination of the truth or falsity of the premises (here the evidence report) is still an empirical matter and thus liable to error. It may thus be seen that confirmation and verification are really only different in degree. While inductive inferences are used in the former and deductive inferences in the latter, the conclusion regarding the hypothesis still must be evaluated in terms of probability. But the reasons that the conclusion in the two cases is limited to a probability statement are different. In confirmation, the probability character of the conclusion results from the probability of the evidence report *and* the fact that an inductive inference must be made. In verification, however, the probability character of the conclusion rests *only* on the probability character of the evidence report since a deductive (not an inductive) inference is made. In general, then, we can say that the conclusion in the case of verification is considerably more probable than that in the case of confirmation. This is so because of the high probability (frequently extremely high) of the evidence report.

Note #3 (from page 251). But this procedure is not quite right. To understand this let us observe that the hypothesis is universal in nature— as we said above, it refers to *all* rats, etc. But we are certainly in no position to test it on all rats; we must be content with a particular sample of rats that we have drawn from a population. We draw a deductive inference that the hypothesis is applicable to our particular sample of rats, i.e., if the hypothesis covers all rats, it certainly covers the particular rats with which we are dealing. By the same token, our evidence report is limited to results obtained on our particular group of rats. But as we have stated, the evidence report (i.e., a and b) is universal in nature. Thus, "a and b" states, in effect, that "*all* rats were reinforced at the end of their maze runs and *all* those rats learned to run the maze with no error. But this is not true. We are able to advance our evidence report for *only* those rats that we studied. Hence, the evidence report is much more specialized or restricted in scope. We should indicate this restriction in some way. This may be accomplished by placing a subscript to the variables contained in the evidence report: a_1 and b_1, which shall then be read "a (certain) group of rats was reinforced at the end of their maze runs and that group of rats learned to run the maze with no errors." Similarly, the hypothesis that is the result of the deductive inference from the general hypothesis must be stated in specialized form: If a_1, then b_1. We then confront the specialized hypothesis with the (specialized) evidence report and reach a conclusion (by way of an inductive inference) as to its truth or falsity. And from this conclusion we go back to our universal hypothesis, again by way of an inductive inference, to determine its probability. This general procedure may thus be represented as follows:

Universal Hypothesis: If a then b
       ↓   (Deductive Inference)
Specialized Hypothesis: If a_1 then b_1
   Evidence Report: a_1 and b_1
       ↓   (Deductive Inference)
     Conclusion: "If a_1 then b_1" is true.
       ↓   (Inductive Inference)
      "If a then b" is probably true.

In terms of our example, the steps are:

Universal Hypothesis:	If (all) rats are reinforced at the end of their maze runs, then those rats will learn to run that maze with no errors.
Specialized Hypothesis:	If a (specific) group of rats is reinforced at the end of their maze runs, then *those* rats will learn to run *that* maze with no errors.
Evidence Report:	A (specific) group of rats was reinforced at the end of their maze runs and *those* rats learned to run that maze with no errors.
Conclusion:	a) The specialized hypothesis is true. b) The universal hypothesis is probably true.

Note #4 (from p. 252). The probability character of empirical hypotheses exerts itself in still other ways, even though they are verifiable, as discussed above. For instance, we may ask how many trials a rat should be run before we are convinced that he would never learn to run the maze perfectly. Say that after a goodly number of trials the rat has still failed to learn. Yet it may be on just the next trial (no matter where we stopped running him) that he would demonstrate flawless performance. One answer is to be more precise in the statement of our antecedent conditions, i.e., to specify a certain number of trials. For example, "If rats are reinforced at the end of each of 30 maze runs, then. . . ." In this case, if they demonstrate the required performance or if they don't after 30 trials, our conclusion is "certain" and the verifiable nature of the hypothesis is preserved.

12

The Inductive Schema—An Overview of Some Characteristics of Science

"Dr. Watson, Sherlock Holmes," said Stamford introducing us.

"How are you?" he said cordially, gripping my hand with a strength for which I should hardly have given him credit. "You have been in Afghanistan, I perceive."

"How on earth did you know that?" I asked in astonishment . . . "You were told, no doubt."

"Nothing of the sort. I knew you came from Afghanistan. From long habit the train of thoughts ran so swiftly through my mind that I arrived at the conclusion without being conscious of intermediate steps. There were such steps, however. The train of reasoning ran, 'Here is a gentleman of a medical type, but with the air of a military man. Clearly an army doctor, then. He has just come from the tropics, for his face is dark, and that is not the natural tint of his skin, for his wrists are fair. He has undergone hardship and sickness, as his haggard face says clearly. His left arm has been injured. He holds it in a stiff and unnatural manner. Where in the tropics could an English army doctor have seen so much hardship and had his arm wounded? Clearly in Afghanistan.' The whole train of thought did not occupy a second. I then remarked that you came from Afghanistan, and you were astonished." (Doyle, 1938, p. 6, 14.)[1]

This, the first meeting between Holmes and Watson is a relatively simple demonstration of Holmes' ability to reach conclusions that confound and amaze Watson. It serves well to illustrate what Reichenbach has called the *inductive schema*. The reconstruction of Holmes' reasoning is presented in the inductive schema shown in Figure 12.1. The observational information available to Holmes is presented at the bottom of that schema. On the basis of this information Holmes infers certain intermediate conclusions. For example, he observed that Watson's face was dark, but that his wrists were fair. These two bits of information immediately led to the conclusion that Watson's skin is not naturally dark. He must therefore have recently been in an area where there was considerable sun—Watson had probably "just come from the tropics."

[1] Reprinted by permission of the Estate of Sir Arthur Conan Doyle.

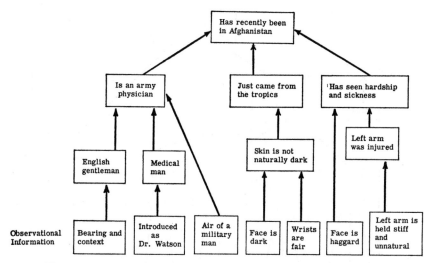

Figure 12.1. An inductive schema based on Sherlock Holmes' first meeting with Dr. Watson.

From the several intermediate conclusions it was then possible for Holmes to infer the final conclusion, that Watson had just recently been in Afghanistan. You should trace through each step of Holmes' reasoning process, as represented in the inductive schema, to make sure that you understand how it was constructed. You might even want to construct such a schema for yourself from any of Holmes' other amazing processes of reasoning.

Let us now turn to an example from physics, an inductive schema that represents the process of scientific reasoning. In Figure 12.2 we have partially reconstructed the evolution of this science. In the bottom row are some of the basic data (evidence reports of investigation) from which more general statements were made.

For instance, Galileo conducted some experiments in which he rolled balls down inclined planes. He measured two variables, the time that the bodies were in motion and the distance covered at the end of various periods of time. He found that the distance traveled was related in a specific manner to the amount of time that the bodies were in motion. This relationship is known as the Law of Falling Bodies.[2]

Copernicus was dissatisfied with the Ptolemaic theory that the sun rotated around the earth, and on the basis of extensive observations and

[2] More precisely, the Law of Falling Bodies is that $S = 1/2\ gt^2$ where S is the distance the body falls, g the gravitational constant, and t the time that it is in motion. History is somewhat unclear about whether Galileo conducted similar experiments in other situations, but it is said that he also dropped various objects off the leaning tower of Pisa and obtained similar measurements.

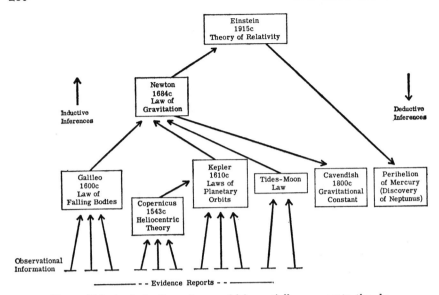

Figure 12.2. An inductive schema which partially represents the development of physics. (After Reichenbach.)

considerable reasoning advanced the Heliocentric (Copernican) Theory of Planetary Motion that the planets rotate around the sun. Kepler based his laws on his own meticulous observations, the observations of others, and on Copernicus' theory. The statement of his three laws of Planetary Orbits (among which was the statement that the earth's orbit is an ellipse) was a considerable advance in our knowledge.

There has always been interest in the height of the tides at various localities, and it is natural that precise recordings of this phenomenon would have been made at various times during the day. Similar observations were made of the location of the moon. Then, on the basis of these two sets of observations, it was possible to state a relationship, known as the Tides-Moon Law. This relationship states that high tides occur only on the parts of the earth that are nearest to, and farthest from, the moon, respectively. It follows that as the moon moves about the earth, the location of high tides shifts accordingly.

Largely on the basis of the preceding relationships, Newton was able to formulate his law of gravitation. Briefly, this law states that the force of attraction between two bodies varies inversely with the square of the distance between them. As an example of a prediction from a general law, we illustrate (the first downward arrow of Figure 12.2) the prediction of the gravitational constant from Newton's law, the precise determination of which was made by Cavendish.

The crowning achievement of the portion of physics that we are considering came with Einstein's statement of his general theory of relativity. One particularly interesting prediction made from the theory of relativity concerned the perihelion of Mercury. Newton's equations failed to account for a slight discrepancy in Mercury's perihelion. This discrepancy between Newton's equations and the observational findings was accounted for precisely by Einstein's theory. The "chain reaction" of discoveries in science is illustrated by the fact that Einstein's work on the movement of Mercury's perihelion led to the discovery of the planet Neptunus by Leverrier.

This brief discussion of the evolution of a portion of physics is, of course, inadequate for a proper understanding of the subject matter involved. Each step in the story constitutes an exciting tale that you might wish to follow up in detail. And where does the story go from here? One of the problems that has been bothering physicists and philosophers is how to reconcile the area of physics depicted in Figure 12.2 with a similar area known as quantum mechanics. To this end physicists such as Einstein and Schrödinger have been attempting to develop a "unified field" theory that will encompass Einstein's theory of relativity as well as the principles of quantum mechanics. Consideration of such higher-level principles is beyond our scope. Our purpose is fulfilled by presenting the general *form* of scientific progress, as shown in Figure 12.2, which is an evolution that may easily be man's crowning intellectual achievement.

We shall now use these two schemata as a basis for considering a number of characteristics of science. Specifically, they will help us (1) to discuss the process of scientific generalization; (2) to elaborate our distinction between inductive and deductive inferences; (3) to understand Reichenbach's concept of concatenation; (4) to understand the nature of explanation, and (5) predictions.

Generalization

Galileo conducted a number of specific experiments. Each experiment resulted in a statement that there was a relationship between the distance traveled by balls rolling down an inclined plane and the time that they were in motion. From these specific statements he then advanced to a more general statement: the relationship between distance and time obtained for the bodies in motion was true for *all* falling bodies, at all locations and at all times.

Copernicus observed the position of the planets relative to the sun. After making a number of specific observations, he was willing to generalize to positions of the planets that he had not observed. The observations that he made fitted the heliocentric theory, that the planets revolved around the sun. He then made the statement that the helio-centric theory held for positions of the planets that he had not observed. And so it is for Kepler's laws and for the Tides-Moon law. In each case a number of specific statements based on observation (evidence reports) were made. Then from these specific statements came a more general statement. It is this process of proceeding from a set of specific state-ments to a more general statement that is referred to as *generalization.* The general statement, then, includes not only the specific statements that led to it but also a wide variety of other phenomena that have not been observed.

This process of increasing generalization continues as we read up the inductive schema. Thus Newton's law of gravitation is more general than any of those that are lower in the schema. We may say that it generalizes Galileo's, Copernicus', Kepler's, and the Tides-Moon Laws. Newton's law is more general in the sense that it includes these more specific laws, and that it makes statements about phenomena that are other than the ones on which it was based. In turn, Einstein formulated principles that were more general than Newton's, principles that in-cluded Newton's and therefore all of those lower in the schema.

More on Inductive and Deductive Inferences

In the inductive schema of Figure 12.2, we may observe that inductive inferences are represented when arrows point up, deductive inferences by arrows that point down. In Figure 12.1 we have only inductive inferences, and these are liable to error. For instance, Watson was introduced as "*Dr.* Watson"; on the basis of this information Holmes concludes that Watson is a medical man. Is this necessarily the case? Obviously not, for Watson may have been some other kind of doctor, such as a Doctor of Philosophy. Similarly, consider the observational information, "left hand held stiff and unnatural," on the basis of which Holmes concluded that "the left arm was injured." This conclusion does not necessarily follow, since there could be other reasons for the condition (Watson might have been organically deformed at birth). In fact, was it necessarily the case that Watson had just come from Afghanistan? The story may well have gone something like this: Holmes: "You have been in Afghan-istan, I perceive." Watson: "Certainly not. I have not been out of London for forty years. Are you out of your mind?"

In a similar vein we may note that Galileo's law was advanced as a general law, asserting that *any* falling body *anywhere* at *any time* obeyed his law. Is this necessarily true? Obviously not, for perhaps a stone falling off Mount Everest or a hat falling off of a man's head in New York, may fall according to a different law than that offered for a set of balls rolling down an inclined plane in Italy many years ago. (We would assume that Galileo's limiting conditions such as that concerning the resistance of air would not be ignored.)

And so it is with the other statements in Figure 12.2. Each conclusion may be in error. As long as inductive inferences are used in such situations, the conclusion will only have a certain degree of probability of being true. Yet, we must continue using inductive inferences. You have no doubt noted that each time a generalization is made, inductive inferences have been used to arrive at that generalization. Since the making of a generalization necessitates saying something about as yet unobserved phenomena, the generalization *must* be susceptible to error.

Let us illustrate deductive logic by referring to our inductive schema for physics. Since Galileo's and Kepler's laws were generalized by Newton's, it follows that they may be deduced from it. In this case, it may be said "If Newton's law is true, then it is necessarily the case that Galileo's is true, and also that Kepler's are true." Similarly, on the basis of Newton's law, the gravitational constant was deduced and empirically verified by Cavendish. This deductive inference takes the form: "If Newton's law is true, then the gravitational constant is such and such." Furthermore, concerning Einstein's principles, we may say: "If Einstein's theory is true, then the previous discrepancy in the perihelion of Mercury may be accounted for."

This discussion brings us to a very important matter which we have only mentioned previously. That is, that a deductive inference does *not* guarantee us that the conclusion is true. The deductive inference discussed above, for example, does not say that Galileo's law is true. It does say that *if* Newton's law is true, Galileo's law is true. One may well ask, at this point, how we determine that Newton's law is true. Or, more generally, how we determine that the premises of a deductive inference are true. The answer, of course, is through the use of inductive logic. For example, we have determined by empirical investigation that the probability of Newton's law being true is very high. It is, in fact, sufficiently high that we wish to say that it *is* true (in an approximate sense, of course).

Concatenation

As we move up the inductive schema we arrive at statements that are increasingly general. And as the generality of a statement increases in the manner depicted in our inductive schema, there is a certain increase in the probability of the statement being true. This increase in probability is the result of two factors. *First*, since the more general statement rests on a wider variety of evidence, it usually has been confirmed to a greater degree than has a less general statement. For example, there is a certain addition to the probability of Newton's law of gravitation that is not present for Galileo's law of falling bodies, since the former is based on inductions from a wider scope of phenomena. *Second,* the more general statement is *concatenated* with other general statements. By concatenated we mean that the statement is "chained together" with other statements, and is thus based on these other statements. For example, Galileo's law of falling bodies is not concatenated with other statements, while Newton's is. The fact that Newton's law is linked with other statements gives it an increment of probability that can not be said of Galileo's. We may say that the probability of the whole system being true is greater than the sum of the probabilities of each statement taken separately. It is the compatibility of the whole system, and the support gained from the concatenation, that provides the added likelihood.

It also follows that when each individual generalization in the system is confirmed, the entire system gains increased credence. For instance, if Einstein's theory was based entirely on his own observations, and those which it stimulated, its probability would be much lower than it actually is, considering that it is also based on all of the lower generalizations in Figure 12.2. Or let us put the matter another way. Suppose that a new and extensive test determined that Galileo's laws were false. This would mean the complete "downfall" of Galileo's laws, but it would only slightly reduce the probability of Einstein's theory since there is a wide variety of additional confirming data for the latter.

Explanation

The concept of explanation as used in science is somewhat difficult for students to understand, probably because of the common-sense use of the term to which they have previously been exposed. One of the common-sense "meanings" of the term concerns familiarity. Suppose that you learn about a scientific phenomenon that is new to you. You want it explained—you want to know "why" it is so. This desire on your part is a psychological phenomenon, a motive. When somebody can relate the scientific phenomenon to something that is already familiar to you, your psychological motive is satisfied. You feel as if you understand the phenomenon because of its association with knowledge that is familiar to you. A metaphor is frequently used for this purpose. For example, it might be said that the splitting of an atom is like shooting a bullet into a bag of sawdust.

Any satisfaction of your motive to relate a new phenomenon to a familiar phenomenon is far from an explanation of it. Explanation is the placing of a statement within the context of a more general statement. If we are able to show that a specific statement belongs in the category of a more general statement, we may say that the specific statement has been explained. To establish this relationship we must show that the specific statement may be logically deduced from the more general statement. To return to a previous example, we might ask how to explain the statement that "John Jones is anxious." The answer is that this statement can be logically deduced from the more general statement that "All men are anxious." This immediately brings us face to face with a matter that we have approached previously from different angles: that such an explanation is accomplished on the assumption that the more general statement is true. Hence, to be more complete, we need to say: "If it is true that 'all men are anxious,' and if it is true that 'John Jones is a man,' then it is true that 'John Jones is anxious.'" By so deductively inferring this conclusion, we have explained why John Jones is anxious— we have logically deduced that specific statement from the more general statement. Of course, we immediately want to go on: why are all men anxious? But this is outside our scope, except that we may note that such an explanation would be accomplished by deducing that statement from a still more general one.

Referring to Figure 12.2 we can see that Kepler's laws are more general than the Copernican theory. And since the latter is included in the former, it may be logically deduced from it—Kepler's laws explain the Copernican theory. In turn, Newton's law, being more general than Galileo's, Kepler's, and the Tides-Moon laws, explains these more specific laws—they may all be logically deduced from Newton's law. And finally, all of the lower generalizations may be deduced from Einstein's theory, and we may therefore say that Einstein's theory explains all of the lower generalizations.

Prediction

To make a prediction we apply a generalization to a situation which has not yet been studied. The generalization says that all of something has a certain characteristic. When we extend the generalization to the new situation, we are simply saying that the new situation should have the characteristic specified in the generalization. In its simplest form this is what a prediction is, and we have illustrated two predictions in Figure 12.2, the gravitational constant and the perihelion of Mercury. Whether or not the prediction is confirmed, of course, is quite important for the generalization. For if the prediction is confirmed, the probability of the generalization is considerably increased. If it is not confirmed, however, (assuming the evidence report is true and that the deduction is valid), then either the probability of the generalization is decreased, or the generalization must be restricted so that it does not apply to the type of phenomena with which the prediction was concerned.

As an illustration of a prediction let us say that a certain hypothesis was formulated about the behavior of school children in the fourth grade. It was then tested on these children and found to be probably true. The experimenter may wonder whether this hypothesis is also true for all school children. If he thinks so he might generalize his hypothesis to make it applicable to *all* school children. From such a generalized hypothesis it is possible to derive specific statements concerning any given school grade. For example, he could deductively derive the conclusion that the hypothesis is applicable to the behavior of school children in the fifth grade, in which case he is making a prediction about children of that grade level. He is, thus, making a prediction that his hypothesis is applicable to a novel situation.

13

Generalization, Explanation, and Prediction in Experimentation

In the last chapter we discussed generalization, explanation and prediction, as well as several other topics, in a rather general way. It now remains for us to consider these topics as they fit into the day to day work of the experimenter. We shall want to discuss some of the mechanics that one might use in these, the final phases of experimentation. The three questions to which we now turn are: How and what does the experimenter generalize? How does he explain his results? And how does he predict to other situations?

Generalization

A distinction is frequently made between what is known as applied science and basic or pure science. In applied science the investigator attempts to solve some relatively limited problem, while in basic science he attempts to arrive at a general principle. The answer that the applied scientist obtains will usually be applicable only under the specific conditions of his experiment. The basic scientist's results, however, are likely to be more widely applicable.

An applied psychologist might be called in by the Burpo Company to find out why their soft-drink sales in Atlanta, Georgia were below normal for the month of December. The basic scientist, on the other hand, would be more likely to study the problem of the general relationship between temperature and consumption of liquids. The applied psychologist, in this example, might find that Atlanta was unseasonably cold during December, and hypothesize that for that reason the people of that city purchased fewer Burpos. The basic scientist, however, might conclude his research with the more general finding that the amount of liquid consumed by humans depends on the temperature—the lower the temperature, the less they consume. Thus, the latter finding would account for the specific phenomenon in Atlanta, as well as a wide variety of additional phenomena—it is by far the more general in scope.

While both kinds of research are important, we shall limit our considerations to basic research. Our immediate goal shall be to see how the experimenter arrives at general statements, rather than only specific statements about the results of his research. We shall start our discussion with the assumption that the experimenter wants to generalize his results as widely as is reasonable. Hence we shall consider two questions in detail: by what procedures does he generalize his results; and how widely is "reasonable"?

The Mechanics of Generalization

Let us say that an experimenter has selected twenty subjects for an experiment. It should be apparent that he is not interested in these twenty subjects in and for themselves, but only insofar as they are typical of a larger group. Whatever he finds out about these subjects he assumes will be true for the larger group. In short, he wishes to *generalize* from his twenty subjects to the larger group of subjects. The terms we have previously used in this connection are "sample" and "population." As we said, an experimenter defines a population of subjects that he wishes to make statements about. This population is usually quite large, such as all the students in the university, all dogs of a certain species, or perhaps even all humans. Since it is not feasible to study all the members of such large populations, the experimenter randomly selects a sample therefrom. And since that sample has been randomly selected from the population, it should be representative of that population. Therefore, the experimenter is able to say that what is probably true for the sample, is also probably true for the population—he generalizes from the sample of subjects to the entire population of subjects from which they came.[1]

The *most important* feature about generalization from a sample, is that the sample must be *representative* of the population. The technique that we are using for obtaining representativeness is randomization—if

---

[1] While this statement offers the general idea, it is not quite accurate. If we were to follow this procedure, we would determine that the mean of a sample is, say, 10.32, and generalize to the population, inferring that its mean is also 10.32. Strictly speaking, this procedure is not reasonable, for it could be shown that the probability of such an inference is .00. A more suitable procedure is known as "confidence interval estimation," whereby one infers that the mean of the population is "close to" that for the sample. Hence, the more appropriate inference might be that, on the basis of a sample mean of 10.32, the population mean is between 10.10 and 10.54.

the sample has been randomly drawn from the population, it is reasonable to assume that it is representative of the population. *Only when the sample is representative of the population are we able to generalize from the sample to the population.* We are emphasizing this point to a great extent for two reasons: because of its great importance in generalizing to populations of *subjects,* and because we ourselves want to state a generalization. We want to generalize from what we have said about subject populations to a wide variety of other populations.[2] For when you conduct an experiment you actually have a number of populations in addition to subjects to which you might generalize.

To illustrate, suppose you are conducting an experiment on knowledge of results. You take two groups of subjects and assign them to two conditions: one group receives knowledge of results, and the second (control) group doesn't. The classic task used in studying this problem is line drawing. Subjects are blindfolded and asked to draw five-inch lines. The knowledge-of-results group would be told whether their lines were too long, too short, or correct, while the control group would be given no knowledge about the lengths of their lines. We are dealing with several populations: of subjects, of experimenters, of tasks, and of various stimulus conditions. Since we wish to generalize to a population of subjects, we randomly draw our sample from that population, and randomly assign them to the two groups. If we find, as we certainly should, that the knowledge-of-results group performs better than the control group, we can safely say that this is probably also true for the population of subjects.

But what about the experimenter? We have controlled this variable, presumably, by having a single experimenter handle all the subjects. But can we say that the knowledge-of-results group will always be superior to the control group *regardless of who is the experimenter?* In short, can we generalize from the results obtained by our single experimenter to all experimenters? This question is difficult to answer. Let us imagine a population of experimenters, made up of all people who conduct experiments. Strictly speaking, then, we should take a random sample from the population of experimenters and have each member of our sample conduct the experiment for himself. Suppose that we define our population of experimenters in such a way that it includes 500 people and that we randomly select a sample of ten experimenters from that population. Further assume that we have selected a sample of 100 subjects. We would then randomly assign the 100 subjects to two groups, then we would randomly assign five subjects in each group to each experimenter. In effect, then, we will repeat our experiment ten times. We

[2] For an elaboration of matters relating to generalization to nonsubject populations you might refer to Brunswick (1956) and Hammond (1948, 1954).

have now not only controlled the experimenter variable by balancing, but also sampled from a population of experimenters. Assume that the results come out approximately the same for each experimenter—that the performance of the knowledge-of-results subjects is about equally superior to their corresponding controls for all ten experimenters. In this case we are able to generalize the results to the population of experimenters as follows: for the population of experimenters sampled (and also for the population of subjects sampled), providing knowledge of results under the conditions of this experiment leads to performance that is superior to that derived from not providing knowledge of results.

By "under the conditions of this experiment" we mean two things: with the specific task used, and under the specific stimulus conditions that were present for the subjects. Concerning the first, our question is this: since we found that the knowledge-of-results group was superior to the control group on a line-drawing task, would that group also be superior in learning other tasks? Of course, the answer is that we do not know. Consider a population of *all* the tasks that human subjects could learn, such as line drawing, learning Morse code, hitting a golf ball, assembling parts of a radio, etc. If we wish to make a statement about the effectiveness of knowledge of results for all tasks, then, as before, we must obtain a representative sample from that population of tasks. By selecting only a line-drawing task we did not do this and therefore cannot generalize back to that population of tasks. The proper procedure to generalize to all tasks would be randomly to select a number of tasks from that population. We would then conduct the same experiment for each of those tasks. If we find that on each task studied the knowledge-of-results group is superior to the control group, then we can say that for *all* tasks, knowledge of results leads to performance that is superior to that gained from a lack of knowledge of results.

Now what about the various stimulus conditions that were present for our subjects? For one, they were blindfolded. But there are different techniques for "blindfolding" subjects. One experimenter might use a large handkerchief, another might use opaque glasses, while still another might place a large screen between the subject's eyes and his hands so that although the subject would be able to see, he could not view the length of his lines. Would the knowledge-of-results condition be superior to the control condition regardless of the technique of blindfolding? What about other stimulus conditions? Would the specific temperature be relevant? How about the noise level? And so on—one can conceive of a number of populations of these stimulus conditions. Strictly speaking, if an experimenter wishes to generalize to the populations of stimuli present, he should randomly sample from those populations. Take temperature as an example. If he wishes to generalize his results to all reasonable values of this variable, then he should randomly select a number of temperatures. He would then repeat his experiment for each temperature value studied. If he finds that regardless of the temperature value studied the knowledge-of-results condition is always superior, he can generalize his findings to the population of temperatures sampled. Only by systematically sampling the various stimulus populations can the experimenter, strictly speaking, generalize his results to those populations.

At this point it might appear that the successful conduct of psychological experimentation is hopelessly complicated. One of the most discouraging features of psychological research is the difficulty encountered in repeating findings. When one Experimenter (X) finds that variable A affects variable B, all too frequently another experimenter (Y) achieves different results. The reason for this lack of repeatability was discussed in Chapter 6, on Control, and Chapter 10, on factorial designs. Looking at it from the present point of view, we might explain the differences in findings by the fact that Experimenter X held a number of conditions constant in his experiment, and then generalized to the populations of these conditions. For example, he may have held the experimenter variable constant, and at least implicitly generalized to a population of experimenters. Strictly speaking, he should not have done that, for he did not randomly sample from a population of experimenters. Let us then assume that his generalization was in error and that the results he obtained are valid only when *he* is the experimenter. If this is the case, then a different set of results may very well be obtained with a different experimenter.

Psychological research (or *any* research for that matter) frequently becomes discouraging. After all, if it were easy there would be little joy to it. The toughest nut to crack yields the tastiest meat. Psychologists must investigate experimental situations more thoroughly than in the past. When the results of two experiments disagree, we should find out what is the matter. This is one of the reasons that we can expect factorial designs to be used more widely in the future, for it is a wonderful device for sampling a number of populations simultaneously. To illustrate, suppose that we wish to generalize our results to populations of subjects, experimenters, tasks, and temperature conditions. We could conduct several experiments here, but let us say that we conduct only one experiment using four independent variables, each varied in the following ways: (1) knowledge of results, two ways (knowledge and no knowledge); (2) experimenters varied in six ways; (3) tasks varied in five ways; and (4) temperature varied in four ways. Assume that we have chosen the values of the last three variables at random. The resulting factorial is presented in Figure 13.1.

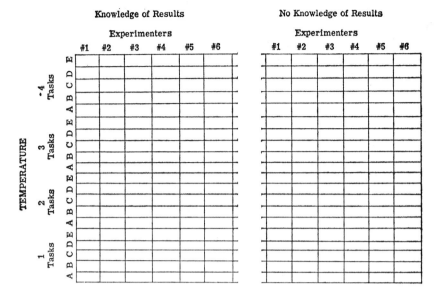

Figure 13.1. A $2 \times 6 \times 5 \times 4$ factorial design for studying the effect of knowledge of results when randomly sampling from populations of experimenters, tasks, temperatures, and subjects.

It can thus be seen that we have a $6 \times 5 \times 4 \times 2$ factorial design. What if we find no difference among all the conditions except that for knowledge of results? This would mean that no variable is affecting performance in a significant manner except the amount of knowledge of results. In this case we could rather safely generalize to our experimenter population, to our task population, to our temperature population, and also, of course, to our subject population.

At this point it is well to recall our discussion from Chapter 10 on factorial designs. There we distinguished between the case of fixed variables and the case of random variables. For the case of fixed variables we said that the experimenter selects the values of his independent variables for some reason—he does not randomly select them from a population. For the case of random variables, however, he defines his population and then randomly selects values from that population. The relevance of that distinction should be apparent, for only in the case of random variables can you safely generalize to the population. If you select the values of your variables in a non-random fashion, any conclusions must be restricted to those values. Let us illustrate by considering the temperature variable again. Suppose that we are particularly interested in three specific values of this variable, 60 degrees, 70 degrees, and 80 degrees. Now, whatever our results, they will be limited to those particular temperature values. On the other hand, if we are interested in generalizing to all temperatures between 40 and 105, we would write each number between 40 and 105 on a piece of paper, place all these numbers in a hat, and draw several values from the hat. Then whatever the experimental results we obtain, we can safely generalize back to that population of values, for we have randomly selected our values from it.[3]

The Limitation of Generalizations

How widely is it reasonable to generalize? Let's say that we are interested in whether Method A or Method B of learning leads to superior performance. Assume that one experimenter tested these methods on a sample of college students and found Method A to be superior. He unhesitatingly generalizes his results to all college students. Another experimenter becomes interested in the problem and repeats the experiment. He finds that Method B is superior. We wish to resolve the contradiction. After studying the two experiments we may find that the first experimenter was in a woman's college, while the second was in a man's college—a possible reason for the different results is now apparent. The first experimenter thus generalized to a population of male *and* female students without randomly sampling from the former (as also did the second, but without sampling females). We suspect that this generalization is in error. To test our suspicions we design a 2 × 2 factorial experiment in which our first variable is methods of learning, varied in two ways, and our second is sex, varied of course in two ways. We randomly draw a sample of males and females from a college population. Assume that our results come out with the following mean values, where the higher the score, the better the performance (Fig. 13.2).

Graphing these results, we can clearly see that an interaction exists between sex and methods such that females are superior with Method A while males are superior with Method B (Figure 13.3). To what extent

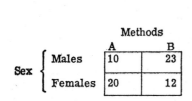

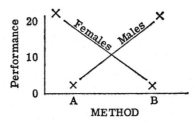

Figure 13.2. A 2 × 2 factorial design with fictitious means.

Figure 13.3. Indicating an interaction between methods of learning and sex.

[3] Assuming of course that we select enough values to study. Just as with sampling from a subject population, the larger the number of values selected, the more likely that the sample is representative of the population.

can we generalize from these findings? Clearly, we cannot say that either Method A or Method B is superior for our population of mixed gender. Rather, we must restrict our generalization. What we can do is generalize to our population of males separately, and to our population of females separately—males of our population will perform better with Method B, females with Method A. The discovery of an interaction, then, limits the extent of our generalization. In general, we can say: that you can generalize your results to a population *unless* the values or characteristics of that population interact with the independent variable.[4] Since this is an important point, let us enlarge on it. In our example of knowledge of results and temperature, we said that if the knowledge-of-results condition is about equally superior for all values of temperature sampled, then we may generalize to that population of temperatures. In this case no interaction exists between temperature and knowledge of results. But suppose that we found such an interaction. For instance, suppose that we found that subjects who receive knowledge of results are superior for temperatures above 60 degrees, but that there is no difference between them and the control subjects when the temperature is lower than 60 degrees. In this event we clearly have to limit our generalization as follows: knowledge of results is superior to no knowledge of results for temperatures above 60 degrees, but this is not the case for temperatures below 60 degrees.

Let us look at a typical experiment in which a number of populations are not sampled. Return to our simple two-groups experiment with knowledge of results, for instance. The only independent variable is amount of knowledge of results, one group getting a positive amount and the second group getting none (or at least a minimal amount). All other (extraneous) variables are controlled, a number of them being held constant at a certain value for all subjects. To what extent can we generalize the results? Consider the experimenter and the task variables, for example. We have not randomly sampled from a population of experimenters nor from a population of tasks. Can we generalize to these populations? Strictly speaking, we shouldn't, but most experimenters generalize in this manner, at least implicitly. Shall we say that they are wrong? Well, without any definite knowledge we cannot condemn them. On the other hand, without some definite knowledge we cannot condone this type of behavior either. For if they generalize to a population of experimenters and tasks, their generalization will be justified if there is no interaction between their independent variable (amount of knowledge of results) and either the population of experimenters or the population of tasks. If an interaction exists between amount of knowledge of results and characteristics of the experimenter or task, however, then the gen-

[4] We are assuming that a random model is used (see p. 231).

eralization will be in error. The answer, then, to the question of how widely one should generalize is this:

First, you are in all probability safe if you generalize only to populations that you have adequately sampled.

Second, if you have previous knowledge that no interaction exists between your independent variable and the populations to which you wish to generalize, then your generalization will probably be valid.

Third, if you have no knowledge about interactions between your independent variable and the populations to which you wish to generalize, then you may tentatively offer your generalization. Other experimenters should repeat your experiment in their own laboratories. This implies that the various extraneous variables will assume different values from those they assumed in your experiment (either as the result of intentional control or because they were allowed to randomly vary). If, in the repetitions of your experiment your results are confirmed, it is likely that the populations to which you have generalized do not interact with your independent variable. On the other hand, if repetitions of your experiment by others, with differences in tasks, stimulus conditions, and other factors do not confirm your findings, then there is probably at least one interaction that needs to be discovered. At this point thorough and piecemeal analysis of the differences between your experiment and the repetitions of it needs to take place in order to discover the interactions. Such an analysis might assume the form of a factorial design such as that diagrammed in Figure 13.1.

This discussion leads us to consider an interesting proposal. Some experimenters have suggested that we should use a highly standardized experimental situation for studying particular types of behavior. All experimenters who are studying a given type of behavior should use the same values of a number of extraneous variables—lighting, temperature, noise, and so forth. In this way we can exercise better control, and be more likely to confirm each other's findings. The Skinner box is a good example of an attempt to introduce standardization of extraneous variables into the experiment, for in the Skinner box the lighting is controlled (it is opaque), the noise level is controlled (it is sound deadened), and a variety of other external stimuli are prevented from entering the box. On the other hand, under such highly standardized conditions the extent to which we can generalize our findings would be sharply limited. If we continue to proceed in the direction we are now going, with each experimenter having different values of his extraneous variables, then when experimental findings are confirmed, we can be rather sure that interactions do not exist. And when findings are not confirmed, we know that we have interactions present that limit our generalizations, and hence we have to initiate experimentation in order to discover them.

Regardless of your opinion on these two positions, that in favor of standardization and that opposed, the matter is probably only academic. Whether because scientists cherish their freedom to establish whatever experimental conditions they want, or because of their laziness, it is unlikely that much in the way of standardization will be accomplished in the foreseeable future.

The final matters for us to discuss in this chapter are those of explanation and prediction, particularly as these are relevant to the day-to-day work of the experimenter. First we shall consider explanation, for after that it will be possible to cover the topic of prediction very briefly.

Explanation[5]

The question "why" lies at the heart of scientific investigation.[6] We have seen that in science "why" amounts to asking: According to what general statements (laws) does a particular event occur? Thus by placing the particular event within the context of a more general statement we can say that the particular event is explained. Let us now consider this process in greater detail.

Hemple and Oppenheim (1948) have offered an example wherein a mercury thermometer is rapidly immersed in hot water. A temporary drop of the mercury column occurs, after which the column rises swiftly. Why does this occur? That is, how might we explain it? Since the increase in temperature affects at first only the glass tube of the thermometer, the tube expands and thus provides a larger space for the mercury inside. To fill this larger space, of course, the mercury level drops. But as soon as the increase in heat is conducted through the glass tube and reaches the mercury, the mercury also expands. And since mercury expands more than does glass (i.e., the coefficient of expansion of mercury is greater than that of glass), the level of the mercury rises.

Now this account, as Hemple and Oppenheim point out, consists of two kinds of statements. Some statements indicate certain conditions that exist before the phenomenon to be explained occurs. These conditions may be referred to as *antecedent conditions* and they include the fact that the thermometer consists of a glass tube which is partly filled with mercury, that it is immersed in hot water, and so on. The second kind of statement expresses general laws, an example of which would be a statement about the thermal conductivity of glass. Now the fact that the

[5] Consult Scriven (1959) for an alternative approach to explanation and prediction.
[6] Some authorities hold that we never ask "why" in science, but rather "what" and "how." We do not mean to quibble about these words, for our actual positions probably would not differ to any great extent. If you prefer "what" and "how" to "why," please use them.

phenomenon to be explained can be logically deduced from the general laws with the help of the antecedent conditions constitutes an explanation of that phenomenon. That is, the way in which we may find out that a given phenomenon can be subsumed under a general law is by determining that the former can be deduced (deductively inferred) from the latter. The schema for accomplishing an explanation can be indicated as follows:

Deductive   ⎧ Statement of the general law(s)
inference: ⎩ Statement of the antecedent conditions
       ↳ Description of the phenomenon to be explained

In this example it can be seen that the phenomenon to be explained (the immediate drop of the mercury level, followed by its swift rise) may be logically deduced according to the above schema. To illustrate further the nature of explanation, we might develop an analogy using the familiar syllogism concerning Socrates. Say that the phenomenon to be explained is Socrate's death. In the syllogism the two kinds of statements that we require for an explanation are offered. First, the antecedent condition is that "Socrates is a man." And second, the general law is that "All men are mortal." From these two statements, of course, we can deductively infer that Socrates is mortal.

Deductive   ⎧ General law: All men are mortal
inference: ⎩ Antecedent condition: Socrates is a man
       ↳ Phenomenon to be explained
         (i.e., Why did Socrates die?) : Socrates is mortal

With this general understanding of the nature of explanation, let us now ask where the procedure enters the work of the experimental psychologist.

Assume that an experimenter wishes to test the hypothesis that the higher the anxiety the better the performance on a relatively simple task. He decides to take as his measure of anxiety the scores that subjects receive on the Manifest Anxiety Scale (Taylor, 1953) where the higher the score, the greater the anxiety.

Say that the experimenter decides to vary anxiety in two ways. He must select two groups of subjects, one group composed of individuals who have considerable anxiety, a second group of those with little anxiety. A relatively simple task is constructed for the subjects to learn. The evidence report states, in effect, that the high-anxiety group performed better than did the low-anxiety group. The evidence report is thus positive, and since it is in accord with the hypothesis, we may say that the hypothesis is confirmed. His experiment is completed, his problem is solved. But is it really? While this may be said of the limited problem for which the experiment was conducted, there is still a nagging question—why is his hypothesis "true"? How might it be explained? To answer this question, of course, he must refer to a principle that is more general than his hypothesis. Let us say that he appeals to a principle, from stimulus-response theory, that says that performance is equal to the amount learned times the drive level present. Letting E stand for performance, H (habit strength) for the *learning factor*, and D for drive, the principle may be stated as $E = H \times D$.

Assuming that anxiety is a specific drive, our experimenter's hypothesis can be established as a specific case of this more general principle. For instance, let us say that the high-anxiety group exhibits a drive factor of 80 units. To simplify matters, assume that both groups learned the task equally well, thus causing the learning factor to be the same for both groups. For instance, H might be 0.50 units. In this event the performance (E) factor for the high-drive group is

$$E = 0.50 \times 80 = 40.00$$

Assume that the low-anxiety group exhibits a drive factor of 20 units, in which case its performance factor would be

$$E = 0.50 \times 20 = 10.00$$

Clearly, then, the performance of the high-drive (high-anxiety) group should be superior to the low-drive group, according to this principle. And the principle is quite general in that it ostensibly covers all drives in addition to including a consideration of the learning factor, H. Following our previous schema, then, we have the following situation:

General Law: The higher the drive, the better the performance (i.e., $E = H \times D$).

Antecedent Conditions: Subjects had two levels of drive, they performed a simple task, anxiety is a drive, etc.

Deductive Inference

Phenomenon to be explained: High-anxiety subjects performed a simple task better than did low-anxiety subjects.

Since it would be possible logically to deduce the experimenter's hypothesis (stated as "the phenomenon to be explained") from the stimulus-response principle together with the necessary antecedent conditions, we may say that the hypothesis is explained.

In Chapter 12 we illustrated the ever-continuing search for a higher-level explanation for general statements. In this chapter we have shown how a relatively specific hypothesis about anxiety and performance can be explained by a more general principle about (1) drives in general and

(2) a learning factor (which we ignored because it was not relevant to the present discussion). The next question, obviously, is how to explain the stimulus-response principle. At this point, however, our immediate purpose is accomplished so that we shall conveniently slip off to other topics.

It is important to note that the logical deduction is made on the assumption that the general principle and the antecedent conditions were actually true. Hence, a more cautious statement about our explanation would be this: assuming that (1) the general law is true, and (2) the antecedent conditions obtained, then the phenomenon of interest is explained. But here's the rub: how can we be sure that the general principle is, indeed, true? We can never be *absolutely* sure, for it must always assume a probability value. It might someday turn out that the general principle used to explain a particular phenomenon was actually false. In this case what we accepted as a "true" explanation was in reality no explanation at all. Unfortunately, we can do nothing more with this situation—our explanations must always be of a tentative sort. We must, therefore, always realize that when we explain a phenomenon it is on the assumption that the general principle used in the explanation is true. If the probability of the general principle is high, then we can feel rather safe. We can, however, never feel absolutely secure. This is merely another indication that we have been given a "probabilistic universe" in which to live. And the sooner we learn to accept this fact (in the present context, the sooner we learn to accept the probabilistic nature of our explanations) the better adjusted to reality we will be.

One final thought on the topic of explanation. We have indicated that an explanation is accomplished by logical deduction. But how frequently do psychologists actually explain their phenomena in such a formal manner? How frequently do they actually cite a general law, state their antecedent conditions, and deductively infer their phenomenon from them? The answer, clearly, is that this is done very infrequently. Almost never will you actually find such a formal process being used in the actual conduct of scientific investigations. Rather, much more informal methods of reasoning are substituted. One need not set out on his scientific career armed with books of logical formulae and the like. But he should be familiar with the basic logical processes that one could go through in order to accomplish an explanation. As with several matters that we have previously discussed, such as stating hypotheses and evidence reports, it is not necessary that you follow precisely the rules that we set down. What *is* important, and what we hope you have gained from this discussion, is that you *could* explain a phenomenon in a formal, logical manner if you wanted or needed to.

Prediction

The *processes* of making predictions and offering explanation are precisely the same, so that everything we have said about explanation is applicable to prediction. The only difference between explanation and prediction is that a prediction is made before the phenomenon is obtained, whereas explanation occurs after the phenomenon is observed. In explanation, then, we start with the phenomenon and logically deduce it from a general law and the attendant antecedent conditions. In prediction, on the other hand, we start with the general law and antecedent conditions, and derive our logical consequence. That is, from the general law we infer that a certain phenomenon should occur. We then conduct our experiment to see if it does occur. If it does, then our prediction has been successful. And as we have previously pointed out, this considerably increases the probability of the general law (unless of course the general law already has a very high degree of probability).

To illustrate a possible prediction briefly, say that we are in possession of the general stimulus-response principle that we previously discussed. We might reason thusly: this principle asserts that the higher the drive, the better the performance; anxiety is a specific drive; therefore, we would predict that high-anxiety individuals would perform a given task better than low-anxiety individuals. The conduct of such an experiment would then inform us of the success (or lack of success) of our prediction. Actually, this is precisely what has been done (cf. Spence, Farber, and McFann, 1956).

14

Miscellany

We have attempted to develop logically the major aspects of experimentation throughout the preceding chapters. Starting with the nature of the problem and concluding with prediction of behavior, we have attempted to weave into a coherent pattern the problems and procedures that are important to the experimental psychologist. It was not feasible to include there *all* matters of importance. Many of these will simply have to await your future study, but in this chapter we can take up several topics that did not conveniently fit into our general plan of development.

Concerning Accuracy of the Data Analysis

In one sense, we would like to place this particular section at the beginning of the book, in the boldest type possible. For no matter how much care you give to the other aspects of experimentation, if you are not accurate in your records and statistical analysis, the experiment is worthless. Unfortunately, there are no set rules that anybody can give you to guarantee accuracy. The best that we can do is to offer you some suggestions which, if followed, will reduce the number of errors, and if you are sufficiently vigilant, eliminate them completely.

The first important point concerns "attitude." Students frequently feel that they must record their data and run their statistical analysis only once, and in so doing, they have amazing confidence in the accuracy of their results. Checking is not for them! While it is very nice to believe in one's own perfection, the author has observed a large enough number of students and scientists over a sufficiently long period of time to know that this is just not reasonable behavior. We all make mistakes.

The best attitude for a scientist to take is not that he *might* make a mistake, but that he *will* make a mistake—his only problem is where to find it. Accept this suggestion or not, as you like. But remember this: at least the first few times that you run an analysis, the odds are about 99 to 1 that you will make an error. As you become more experienced, the odds might drop to about 10:1. The author once had occasion to talk with one of our most outstanding statisticians. To decide a matter it became necessary to run a simple statistical test. Our answer was obviously absurd, so we tried to discover the error. After several checks, however, the fault remained obscure. Finally, a third person, who could look at the problem from a fresh point of view, checked our computations and found the error. The statistician admitted that he was never very good in arithmetic, and he frequently made errors in addition and subtraction.

The first place that an error can be made occurs when you first start to obtain your data. More often than not the experimenter observes behavior and records it by writing it down, so let us take such a case as an example.

Suppose that you are running rats in a T-maze, and that you are recording (1) their latency, (2) their running time, and (3) whether they turned left or right. You might take a large piece of paper on which you can identify your subject, and have three columns for your three kinds of data, noting the data for each subject in the appropriate column. But once you indicate the time values and the direction the rat turned, you move on to your next subject; the event is over and there is no possibility for further checking. Hence, any error you make in writing down your data is uncorrectable for all eternity. You should therefore be exceptionally careful in recording the correct value. You might fix the value firmly in mind, and then write it down, asking yourself all the time whether you are transcribing the right value. After it is written down, check yourself again to make sure that it is correct. If you find a value that seems particularly out of line, you might double-check it to see if it is right. After double-checking such an unusual datum, it is worthwhile to make a note that it is correct, for later on you might return to it with considerable doubt. For instance, if most of your rats take about two seconds to run the maze, but if you write down that one rat had a running time of 57 seconds, take an extra look at the clock to make sure that this reading is correct. Then, if it is, make a little note beside "57 seconds," indicating that the value has been checked.

Frequently, experimenters transcribe the original records of behavior onto another sheet for their statistical analysis. Such a job is long and tedious, and therefore conducive to errors. In recopying data onto new sheets, then, considerable vigilance must be exercised. The finished job should be checked to make sure that no errors in transcription have been committed. (It is frequently possible to avoid this step. For instance, if you can plan your data sheet so that you can record the measures of behavior directly on the sheet that you will use for your statistical analysis, you will prevent errors of transcription.)

In writing data on a sheet, legibility is of utmost importance, for the reading of numbers is a frequent source of error. You may be surprised at the difficulty you might have in reading your own writing, particularly after a period of time. If you use a pencil, that pencil should be quite sharp and hard, to reduce smudging. If possible, record your data in ink. And if you have to change a number, it first must be thoroughly erased or eradicated with ink eradicator.

Labeling of all aspects of your data sheet should be complete, since you may wish to refer to the data at some later time. You should label the experiment clearly, giving its title, the date, place of conduct, and so on. You should unambiguously label each source of data. Your three columns might be labeled, for example, "latency of response in leaving start box," "time in running from start box to close of goal box door," and "direction of turn." Each statistical operation should be clearly labeled. If you run a t test, for instance, the top of your work sheet should state that it is a t test between such and such conditions, using such and such a measure as the dependent variable. In short, label everything pertinent to the records and analysis so that you can return to your work years later and understand it readily.

The actual conduct of the statistical analysis is probably going to be the greatest source of error. It is thus advisable to check each step as you move along. For example, you will probably begin by computing the sums and sums of squares for your groups. Before you substitute these values into your equation you should check them. Otherwise, if they are in error, all of your later work will have to be redone. Similarly, each multiplication, division, subtraction and addition should be checked just after it has been made, before you move on to the next operation which incorporates the result. After you have computed your statistical test, checking each step along the way, you should put it aside and do the entire example again, without looking at your previous work. If your second computation, performed independently of the first, checks with your first computation, the probability that you have erred is decreased (it is not eliminated, of course, for you may have made the same error twice).

It will be advantageous to have someone else conduct the same statistical analysis so that your results and his can be compared. It is also advisable to indicate when you have checked a number or operation. One way to accomplish this is to place a small dot above and to the right of the value (do not place it so low that the dot might be confused with a decimal point). The values of indicating a checked result are (1) that you can better keep track of where you are in your work, and (2) that at some later time you will know whether or not the work has been checked.

Concerning the statistical analysis, it might be well to point out a source of many errors. When you are conducting your statistical analysis, it is easy to leave out steps. For instance, if your equation calls for you to square a term and then divide that term by the number of subjects, you might tend to do both of these operations at once, merely writing down the result. If you will try *not* to do this, not only will you find that your errors are reduced, but you will be able to check each step of your work more closely. In the above example, for instance, you should write down the square of the number and its divisor. Then write down the result of the division.

Combining Tests of Significance

Experimenters frequently have available two or more sets of experimental results that test the same hypothesis. Now, since the experiments are independent of each other, it is possible to combine the results of the statistical tests. While this procedure may be used for several reasons, one particularly advantageous one is that in neither of the separate experiments was it possible to reject the null hypothesis. Yet the means of the two groups might have been in the same direction, and their differences sufficiently great so that they were strongly suggestive. In such cases it is possible to combine the tests of significance in order to obtain a sounder test of the empirical hypothesis.

A number of techniques for combining two or more tests of significance are available (Lindquist, 1953; Mosteller and Bush 1954, Chapter 8). While we cannot possibly go into the advantages and disadvantages of each technique, one approach is extremely easy to use, although it is applicable only to the case where there are but two experiments. To illustrate, say that in one experiment the probability that the null hypothesis is false is 0.08, while in the second experiment it is 0.10. Clearly, in neither experiment was it possible to reject the null hypothesis, assuming the significance level was set at 0.05. On the further assumption that the means of the two groups in both experiments were in the same direction (e.g., the experimental group had the higher mean in both cases), we can combine these findings by referring to Figure 14.1. Thus, we locate 0.08 on the horizontal axis, and .10 on the vertical axis (we could of course reverse these if we wished). Reading up and across until the lines for the two values of P intersect, we obtain the combined probability. In this example it is less than 0.05, so that, considering the two experiments as combined, we are able to reject the null hypothesis, whereas considering them separately this was not possible.

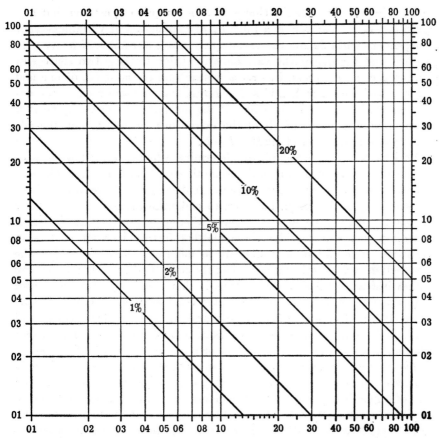

Figure 14.1. The probability of two combined tests of significance. Find the value of one probability along a horizontal axis, and the second probability along a vertical axis. Where the vertical line crosses the horizontal line one finds a diagonal that indicates the combined probability value. Five of the more commonly used significance levels are indicated here. (From P. C. Baker, personal communication.)

Assumptions Underlying the Use of Statistical Tests

In applying the statistical tests to the experimental designs presented in this book, one must make certain assumptions. In general, these are that (1) the dependent variable scores are independent; (2) the variances of the groups are equal (homogeneous); (3) the population distribution is normally distributed; (4) the treatment effects and the error effects are additive. You should get a rough idea of the nature of assumptions 2 and

3. To help you visualize the character of assumption 4, assume that any given dependent variable is a function of two classes of variables—your independent variable and the various extraneous variables. Now we may assume that the dependent variable values due to these two sources of variation can be expressed as an algebraic sum of the effect of one and the effect of the other, i.e., if R is the response measure used as the dependent variable, if I is the effect of the independent variable, and if E is the combined effect of all of the extraneous variables, then the additivity assumption says that $R = I + E$.

Various tests are available to determine whether or not your particular data allow you to regard assumptions 2, 3, and 4 as tenable, and therefore justify your statistical tests. It does not seem feasible, at the present level of approach, however, to elaborate these assumptions nor the nature of the tests for determining whether or not they are satisfied. Certainly you will consider these matters further in your courses in statistics. In addition it is difficult to determine whether or not the assumptions are exactly satisfied, i.e., the tests used for this purpose are rather insensitive. And the consensus is that rather sizeable departures from them can be tolerated, still yielding valid statistical analyses (Lindquist, 1953; McNemar, 1955; Dixon and Massey, 1951; Cochran and Cox, 1957; Anderson and Bancroft, 1952; Boneau, 1960, etc.). For further information, you should consult these or a number of other references.

The first assumption, however, is essential, since each dependent variable score must be *independent of every other dependent variable score.*[1] For example, if one score is 15, the determination that a second score is, say, 10 must in no way be influenced by the fact that the first score is 15. If subjects have been selected at random, and if one and only one score on each dependent variable is used for each subject, then the assumption of independence should be satisfied. Perhaps the most frequent error made in violation of this assumption is illustrated by the following. Suppose that in a learning experiment we conduct, a certain subject yields scores of 10, 8, 6, 5, and 4 on each of five trials. Some students might want to refer to each of these scores as values of X in computing ΣX. Accordingly, they might say that $n = 5$. Now, this is a clear violation of the assumption of independence, for the second score is related to the first score, the third to the first and second, and so on. That is, all of these scores were made by the same subject, and if he happens to be capable, all the scores will tend to be high. A second error that would be committed

---

[1] In the case of the matched groups design, the independence assumption takes a slightly different form, that is that the values of D are independent. Hence, a more adequate statement of this assumption would be that the treatment effects and the error are independent. That is, in terms of the symbols used for the fourth assumption, I and E are independent.

in this procedure, is the artificial inflation of n. That is, these scores are all for one subject; therefore, n cannot possibly equal 5. The proper procedure would be to obtain one single score for this subject, say by adding them up and using the sum as the dependent variable score. Hence, the proper values are $\Sigma X = 33$ and $n = 1$.

We have not emphasized statistical assumptions, except for the one concerning independence, primarily because of the introductory nature of this book. As you progress in statistical and experimental work, you will want to learn more about the assumptions of homogeneity of variances and normality of the populations. With a more thorough knowledge of these assumptions, and how you might determine when they are met, you will be in a better position to evaluate possible errors introduced by violating them. And, of course, when really serious violations of these assumptions occur, you will have to effect some remedy. In general, there are two possible remedies. One is to transform your data (e.g., to take the square root or reciprocal of each score) and continue the analysis in the usual way. The second would be to use a type of statistical test known as a nonparametric test, which does not require that you meet the assumptions of normality and homogeneity of variances. Nonparametric tests still require the assumption of independence, however.

Reducing Error Variance

We have previously discussed the term *error variance* to some extent. Because of its importance in experimentation, we shall consider it in greater detail here.

In every experiment there is a certain error variance, and in our statistical analysis we obtain an estimate of it. For example, in our two-groups designs, the error variance is the denominator of the t ratio. Where we use Duncan's range test, our estimate of the error variance is provided by equation 9.5 (p. 179), the square root of the error variance. And in factorial designs, the error variance is the denominator of the F ratio. Basically, these three estimates of the error variance are measures of the same thing—the extent to which subjects treated alike exhibit variability in their dependent variable scores. There are many reasons why we obtain different scores for subjects treated alike. For one, subjects are all "made" differently, and they will all react somewhat differently to the same experimental treatment. For another, it simply is impossible to treat all subjects in the same group precisely the same; we always have a number of extraneous variables operating on our subjects in a random manner, resulting in differential effects. And finally, some of the error variance is due to imperfections in our measuring devices. No device can provide a completely "true" score, nor can we as humans make completely accurate and consistent readings of the measuring device.

In many ways, it is unfortunate that dependent variable scores for subjects treated alike show so much variability, but we must learn to live with the fact that this variability will always be with us. The best we can do is to attempt to reduce it. The reason that we want to reduce the error variance in an experiment should be apparent, but to make sure that we understand, let us consider an example. We shall henceforth consider the t test, as used in a two-groups design, but what we have to say may be considered applicable to other designs and their statistical tests. Let us say that we know that the difference between the means of two groups is 5. Consider two situations, one where the error variance is rather large, and one where it is rather small. For example, say that the error variance is 5 in the first case, but 2 in the second. For the first case, then, our computed t would be: $t = 5/5 = 1.0$, and for the second it would be $t = 5/2 = 2.5$. Clearly, in the first case we would fail to reject our null hypothesis, while in the second case we are likely to reject it. For both situations we have the same mean difference. Such a difference in our two experiments is, of course, of the utmost importance. We naturally seek to reject the null hypothesis, if it "should" be rejected. What we mean is that the null hypothesis specifies that there is no difference between our two groups. However, in a given experiment we find that the means of our groups differ. The question is whether or not the difference is sufficiently large to allow us to reject the null hypothesis. Now if our error variance is sufficiently large, we shall not. But it may be rejected if the error variance is sufficiently small. Hence we seek to obtain an error variance that is sufficiently small to allow us to reject our null hypothesis. If after reducing it as much as possible, we still cannot reject our null hypothesis, then it seems reasonable to conclude that the null hypothesis should actually not be rejected. Let us now consider ways in which we can reduce the error variance in our experiments.

First, we should be clear that we are *not* trying to find out how to increase our chances of rejecting the null hypothesis. As we have previously pointed out, it is frequently possible to increase your chances of rejecting the null hypothesis by exaggerating the difference in the independent variable values that you administer to your two groups. For instance, if we are seeking to determine whether amount of practice affects amount learned, we are more likely to obtain a significant difference between two groups if we have them practice 100 trials versus 10 trials, than if we have them practice 15 trials versus 10 trials. This is so because we would probably increase the difference between the means on the dependent variable of the two groups, and the greater the mean difference, the larger the value of t. And, of course, the larger the value of t, the more likely it is to be significant. In order to discuss the ways in which the error variance can be reduced, let us get the basic formula for t for the two-randomized-groups design before us:[2]

$$(14.1) \qquad t = \frac{\overline{X}_1 - \overline{X}_2}{\sqrt{s_1^2/n_1 + s_2^2/n_2}}$$

We have already seen that the larger the numerator, the larger the value of t, and the smaller the denominator, the larger the value of t. But let us now consider the denominator in greater detail. First, we can see clearly that as the variances of the groups decrease, the size of t increases. To illustrate, assume that the mean difference is 5, that the variances are each 64 and that n_1 and n_2 are both 8. In this event,

$$t = \frac{5}{\sqrt{64/8 + 64/8}} = 1.25$$

Now let us say that the experiment is conducted again, except that this time we are able to reduce the variances to 16. In this case,

$$t = \frac{5}{\sqrt{16/8 + 16/8}} = 2.50$$

Granting, then, that it is highly advisable to reduce the variance of our groups, how can we accomplish this? There are several possibilities. First, recall that our subjects, when they enter the experimental situation, are all different, and that the larger such differences, the greater the variances of our groups. Therefore, one obvious way to reduce the variances of our groups, and hence the error variance, is to reduce the extent to which our subjects are different. Psychologists frequently increase the homogeneity of their groups by selection. For example, we work with a number of different strains of rats. In any given experiment, however, all the subjects are usually taken from a single strain—the Wistar strain, the Sprague-Dawley strain, or whatever. If a psychologist randomly assigns rats from several different strains to his groups, he is probably going to increase his variances. Working with humans is more difficult, but even here the selection of subjects that are similar is a frequent practice, and should be considered. For example, using college students as subjects undoubtedly results in smaller variances than if we selected subjects at random from the general population of humans. But you could be even more selective in your college population—you might

[2] See footnote 11, p. 153.

use only females, only subjects with IQs above 120, only students with low anxiety scores, and so on.

At this point, one serious objection that comes to mind is that by selection of subjects you thus restrict the extent to which you can generalize your results. Thus, if you sample only high-IQ students, you will certainly be in danger if you try to generalize your findings to low-IQ students, or to any other population that you have not sampled. For this reason, selection of homogeneous subjects should be seriously pondered before it is adopted. Unfortunately, you cannot have your cake and eat it too—the greater the extent to which you select homogeneous subjects, the less you can generalize.[3]

A *second* way in which you can reduce your variances is in your experimental procedure. The ideal is to treat all subjects in the same group as precisely alike as possible. We cannot emphasize this too strongly. We have counseled the use of a tape recorder for administering instructions, in order that all subjects would receive precisely the same words, with precisely the same intonations. If you rather casually tell your subject what to do, varying the way in which you say it with different subjects, your variances are probably increasing. Similarly, the greater the number of extraneous variables that are operating in a random fashion, the greater will be your variances. If, for example, noises are present in varying degrees for some subjects, but not present at all for others, your group variances are probably going to increase. Here again, however, you should recognize that if you eliminate extraneous variables to any great extent, you will have difficulty in generalizing to situations where they are present. For example, if all your subjects are run in sound-deadened rooms, then you should not, strictly speaking, generalize to situations where noises are randomly present. But since we usually are not trying to generalize, at least not immediately, to such uncontrolled stimulus conditions, this general objection need not greatly disturb us.

A *third* way to reduce your variances concerns the errors that you might make—errors in reading your measuring instruments, in recording your data, and in your statistical analysis. The more errors that are present, the larger will be the variances, assuming that such errors are of a random nature.

The three techniques noted above are ways of reducing the error variance by reducing the variances of your groups. Another possible technique for reducing the error variance concerns the design that you select. The

[3] Of course if you systematically sample a population and have a number of different values for the population, you are in a better position to generalize. For instance you might classify high-IQ, medium-IQ, and low-IQ subjects, in which case you would want each group to be homogeneous. The comments in the text apply only when you do not systematically sample along some dimension.

clearest example for the designs that we have considered would be to replace the two-randomized-groups design with the matched-groups design for two groups, providing that there is a substantial correlation between the independent variable and dependent variable. As you may recall, the error variance is reduced in accordance with the size of the correlation (p. 159). The factorial design can also be used to decrease your error variance. For example, you might incorporate an otherwise extraneous variable in your design, and remove the variance attributable to that variable from your error variance. This and similar uses of the factorial design are, however, beyond the scope of this book.

Referring back to equation 14.1, we have seen that as the variances of our groups decrease, the error variance decreases, and the size of t increases. The other factor in the denominator is n. As the size of n increases, the error variance decreases. This occurs for two reasons: because we use n in the computation of the variances, and because n is also used otherwise. We might comment that increasing the number of subjects per group is probably one of the easiest ways to decrease, usually very sharply, the error variance. More will be said shortly about the number of subjects in an experiment.

A technique that is frequently effective in reducing error variance is the "analysis of covariance." Briefly, this technique enables one to obtain a measure of what you think is a particularly relevant extraneous variable that you are not controlling. This usually has to do with some characteristic of your subjects. For instance, if you are conducting a study of the effect of certain psychological variables on weight, you might use as your measure the weight of your subjects before you administer your experimental treatments. Through analysis of covariance, you then can "statistically control" this variable—that is, you can remove the effect of initial weight from your dependent variable scores, thus decreasing your error variance. We might note that the degree of success in reducing error variance with the analysis of covariance depends on the size of the correlation between your extraneous variable and your dependent variable. In your future study of experimentation and statistics you should attempt to learn how this technique is applied.

We have tried in this section to indicate the importance of the reduction of error variance in experimentation, and to suggest some of the ways that it might be accomplished. Unfortunately, it is not possible to provide an exhaustive coverage of the available techniques, because of both lack of space and complexities that would take us beyond our present level of discussion. Excellent treatments of this topic have been given elsewhere, although they require a somewhat advanced knowledge of experimentation and statistics (e.g., Cochran and Cox, 1957; Fisher, 1953).

Number of Subjects per Group

"How many subjects should I have in my groups?" is a question that students usually ask in a beginning course in experimental psychology. Experienced experimenters, however, seem never to ask this question, largely because in that form it cannot be answered. Experience has taught them that they must "play it by ear." They usually run a number of subjects, more or less arbitrarily determined, and see how the results turn out. If their groups differ significantly, they may be satisfied with the numbers they chose, or they may run additional subjects to confirm their significant findings. On the other hand, if their groups do not differ significantly, but the differences are promising, they may run more subjects in the hope that the additional data will produce significance.[4]

While we cannot adequately answer the student's question, we can offer some guiding considerations. First, the larger the number of subjects run, the more reliably can be estimated the difference (if such exists) between the groups. This is a true and sure statement, but it does not help very much. We can clearly say that 100 subjects per group is better than 50. You may want to know if 20 subjects per group is enough. That depends, first of all on the "true" (population) difference between your groups, and second on the size of the variances of your groups. What we can say is that the larger the true difference between groups, the smaller the number of subjects required for the experiment; and the smaller the group variances, the fewer subjects required. Now if you know what the differences are, and also what the variances are, the number of subjects required can be estimated. Unfortunately, experimenters do not usually have this information, or if they have it they do not consider the matter worth the effort required to answer the question. We shall not attempt to judge what should or should not be done in this respect, but shall illustrate the procedure for determining the minimum number of subjects required, given these two bits of information. (Possible sources of this information include (1) an experiment reported in the literature similar to the one you want to run, from which you can abstract the necessary information and (2) a pilot study conducted by yourself to yield estimates of the information that is needed.)

[4] This latter procedure cannot be legitimately (strictly speaking) defended in other than a preliminary investigation. Clearly, one who keeps running subjects until he obtains a significant difference may capitalize on chance. For example, if one runs 10 subjects per group and obtains a t value that approaches significance, he might run 10 more subjects per group. Assume that the t is now significant. But the results of these additional subjects might be due merely to chance. The experiment is stopped and success proclaimed. If still more subjects were run, however, significance would be lost, and the experimenter would never know this fact. If such an experiment is to be cross-validated, this procedure is, of course, legitimate.

In any event, let us suppose that you are going to conduct a two-randomized-groups experiment. You estimate that the mean score of Condition A is 10, and that the mean of Condition B is 15. The difference between these means is 5. You also estimate that the variances of your two groups are both 75. Say that you set your significance level at .05, in which case the value of t that you will need to reject the null hypothesis is *approximately* 2 (you may be more precise if you like). Assume that you want an equal number of subjects in both groups. Now we have this information:

$$\overline{X}_1 - \overline{X}_2 = 5$$
$$s_1^2 \text{ and } s_2^2 \text{ both } = 75$$
$$t = 2$$

Let us solve equation 14.1 for n instead of for t. By simple algebraic manipulation we find that, on the above assumptions, equation 14.1 becomes

(14.2)
$$n = \frac{2t^2s^2}{(\overline{X}_1 - \overline{X}_2)^2}$$

Substituting the above values in equation 14.2 and solving for n, we find

$$n = \frac{2(2)^2(75)}{(15 - 10)^2} = \frac{600}{25} = 24$$

We can say, therefore, that with this true mean difference, and with these variances for our two groups, and using the 5 per cent level of significance, we need a minimum of 24 subjects per group to achieve significance. We have only approximated the value of t necessary at the 5 per cent level, however, and we have not allowed for any possible increase in the variance of our two groups. Therefore, we should underline the word *minimum*. To be safe, then, we should probably run somewhat more than 24 subjects per group—30 would seem reasonable in this case.[5]

We have illustrated a technique for estimating the minimum number of subjects necessary to reject the null hypothesis, if in fact you can reject it, for the case of the two-randomized-groups design. The technique can also be extended, with a little industry, to the equations for analyzing the other designs that we have taken up.

---

[5] This procedure is offered only as a rough guide, for we are neglecting power considerations of the statistical test. This procedure has a minimal power for rejecting the null hypothesis.

References

Anastasi, A. The place of experimental psychology in the undergraduate curriculum. *Amer. Psychol.*, 1947, 2, 57-62.

Anderson, R. L. and Bancroft, T. A. *Statistical theory in research.* New York: McGraw-Hill, 1952.

Asher, R. Why are medical journals so dull? *Brit. Med. J.*, 1958, II, 502.

Bechtoldt, H. P. Construct validity: a critique. *Amer. Psychol.*, 1959, 14, 619-629.

Boneau, C. A. "The effects of violation of assumptions underlying the *t* text." *Psychol. Bull.* 1960, 57, 49-64.

Bridgeman, P. W. *The logic of modern physics.* New York: Macmillan, 1927.

Brown, C. C. & Saucer, R. T. *Electronic instrumentation for the behavioral sciences.* Springfield, Illinois: Charles C. Thomas, 1958.

Brunswik, E. *Perception and the representative design of psychological experiments.* Berkeley and Los Angeles: Univer. Calif. Press, 1956.

Bugelski, B. R. *Experimental psychology.* New York: Henry Holt, 1951.

Buxton, C. E. *College teaching: A psychologist's view.* New York: Harcourt, Brace, 1956.

Cantril, H. *The invasion from Mars.* Princeton: Princeton Univer. Press, 1940.

Cochran, W. G. & Cox, G. M. *Experimental designs.* New York: Wiley, 1957.

Cohen, M. R. & Nagel, E. *Logic and scientific method.* New York: Harcourt, Brace, 1934.

Cook, B. S. and Hilgard, E. R. Distributed practice in motor learning: Progressively increasing and decreasing rests. *J. exp. Psychol.*, 1949, 39, 169-172.

Cronbach, L. J. & Meehl, P. E. Construct validity in psychological tests. *Psychol. Bull.*, 1955, 52, 281-302.

Deese, J. *The psychology of learning.* New York: McGraw-Hill, 1952.

Dixon, W. J. & Massey, F. J. Jr. *Statistical analysis.* New York: McGraw-Hill, 1951.

Doré, L. R. and Hilgard, E. R. Spaced practice as a test of Snoddy's two processes in mental growth. *J. exp. Psychol.*, 1938, 23, 359-374.

Doyle, A. C. *Sherlock Holmes.* Garden City, New York: Garden City Press, 1938.

Dubs, H. H. *Rational induction.* Chicago: Univer. Chicago Press, 1930.

Duncan, D. B. Multiple range and multiple *F* tests. *Biometrics*, 1955, 11, 1-42.

Duncan, D. B. Multiple range tests for correlated and heteroscedastic means. *Biometrics*, 1957, 13, 164-176.

Duncan, D. B. A simple Bayes solution to a common multiple comparisons problem. *Institute of Statistics Mimeograph Series No. 223, Univer. North Carolina,* April 1959.

Edwards, A. L. *Experimental design in psychological research.* New York: Rinehart, 1950.

Feigle, H. and Scriven, M. *Minnesota studies in the philosophy of science.* (Volume I: The foundations of science and the concepts of psychology and psychoanalysis.) Minneapolis: Univer. Minnesota Press, 1956.

Fisher, R. A. *The design of experiments* (6th ed.). New York: Hafner, 1953.

Frank, P. G. (ed.) *The validation of scientific theories.* Boston: Beacon, 1956.

Gaito, J. Statistical dangers involved in counterbalancing. *Psychol. Rep.,* 1958, 4, 463-468.

———, Multiple comparisons in analysis of variance. *Psychol. Bull.,* 1959, 56.

Grings, W. W. *Laboratory instrumentation in psychology.* Palo Alto, California: National Press, 1954.

Guilford, J. P. *Fundamental statistics in psychology and education.* (3rd ed.). New York: McGraw-Hill, 1956.

Hammond, K. R. Subject and object sampling—a note. *Psychol. Bull.,* 1948, 45, 530-533.

Hammond, K. R. Representative vs. systematic design in clinical psychology. *Psychol. Bull.,* 1954, 51, 150-159.

Hempel, C. G. Studies in the logic of confirmation. *Mind,* N. S., 1945, 54, 1-26, 97-121.

Hempel, C. G. and Oppenheim, P. The logic of explanation. *Phil. Sci.,* 1948, 15, 135-175.

Hepler, J. W. On the teaching of experimental psychology. *Amer. Psychol.,* 1959, 14, 638-641.

Hull, C. L. *Principles of behavior.* New York: Appleton-Century, 1943.

Hull, C. L. *A behavior system.* New Haven: Yale Univer. Press, 1952.

Jenkins, J. G. and Dallenbach, K. M. Oblivescence during sleep and waking. *Amer. J. Psychol.,* 1924, 35, 605-612.

Kramer, C. Y. Extension of multiple range tests to group means with unequal numbers of replications. *Biometrics,* 1956, 12, 307-310.

Kramer, C. Y. Extension of multiple range tests to group correlated adjusted means. *Biometrics,* 1957, 13, 13-18.

Li, J. C. R. *Introduction to statistical inference.* Ann Arbor, Michigan: Edwards Brothers, 1957.

Lindquist, E. F. *Design and analysis of experiments in psychology and education.* Boston: Houghton Mifflin, 1953.

Loevinger, J. Objective tests as instruments of psychological theory. *Psychol. Rep.,* 1957, 3, 635-694.

McGuigan, F. J. and MacCaslin, E. F. Whole and part methods in learning a perceptual motor skills. *Amer. J. Psychol.,* 1955, 47, 658-661. (a)

McGuigan, F. J. & MacCaslin, E. F. The relationship between rifle steadiness and rifle marksmanship and the effect of rifle training on rifle steadiness. *J. appl. Psychol.*, 1955, 39, 156-159. (b)

McGuigan, F. J. Confirmation of theories in psychology. *Psychol. Rev.*, 1956, 63, 98-104.

McKeachie, W. J., Pollie, D. & Speisman, J. Relieving anxiety in classroom examinations. *J. abnorm. soc. Psychol.*, 1955, 50, 93-98.

McNemar, Q. *Psychological statistics* (2nd ed.). New York: Wiley, 1955.

Minami, H. and Dallenbach, K. M. The effect of activity upon learning and retention in the cockroach, *Periplaneta americana*. *Amer. J. Psychol.*, 1946, 59, 1-58.

Morgan, C. L. *An introduction to comparative psychology*. (2nd ed.). London: Walter Scott, 1906.

Mosteller, F. & Bush, R. R. Selected quantitative techniques. In G. Lindzey (Ed.) *Handbook of Social Psychology*. Cambridge, Mass.: Addison-Wesley, 1954.

Mowrer, O. H. *Psychotherapy*. New York: Ronald, 1953.

Newbury, E. Current interpretation and significance of Lloyd Morgan's canon. *Psychol. Bull.*, 1954, 51, 70-74.

Page, I. H. Serotonin. *Sci. Amer.*, 1957, 197, 52-56.

Peters, C. C. & Van Voorhis, W. R. *Statistical procedures and their mathematical bases*. New York: McGraw-Hill, 1940.

Reichenbach, H. *Elements of symbolic logic*. New York: Macmillan, 1947.

Reichenbach, H. *Experience and prediction*. Chicago: Univer. Chicago Press, 1938.

Reichenbach, H. *The theory of probability* (2nd ed.). Berkeley and Los Angeles: Univer. Calif. Press, 1949.

Renshaw, S. and Schwarzbeck, W. C. The dependence of the form of the pursuit meter learning function on the length of the inter-practice rests: I. Experimental. *J. gen. Psychol.*, 1938, 18, 3-16.

Ryan, T. A. Multiple comparisons in psychological research. *Psychol. Bull.*, 1959, 56, 26-47.

Sandler, J. A test of the significance of the difference between the means of correlated measures, based on a simplification of Student's *t*. *Brit. J. Psychol.*, 1955, 46, 225-226.

Scriven, M. Explanation and prediction in evolutionary theory. *Sci.*, 1959, 130, 477-482.

Silverman, H. F. Effects of praise and reproof on reading growth in a non-laboratory classroom setting. *J. educ. Psychol.*, 1957, 48, 199-206.

Skinner, B. F. *Science and human behavior*. New York: Macmillan, 1953.

Spence, K. W. The postulates and methods of 'Behaviorism.' *Psychol. Rev.*, 1948, 55, 67-78.

Spence, K. W., Farber, I. E. and McFann, H. H. The relation of anxiety (drive) level to performance in competitional and noncompetitional paired-associates learning. *J. exp. Psychol.*, 1956, 52, 296-305.

Taylor, J. A. A personality scale of manifest anxiety. *J. abnorm. soc. Psychol.*, 1953, 48, 285-290.

Underwood, B. J. *Experimental Psychology.* New York: Appleton-Century-Crofts, 1949.

Underwood, B. J., Duncan, C. P., Taylor, J. A. & Cotton, J. W. *Elementary Statistics.* New York: Appleton-Century-Crofts, 1954.

Underwood, B. J. *Psychological research.* New York: Appleton-Century-Crofts, 1957.

Underwood, B. J. Training for research in experimental psychology. *Psi Chi Newsletter,* January 1958.

Woodworth, R. S. & Schlosberg, H. *Experimental psychology.* New York: Henry Holt, 1955.

Powers and Roots

No.	Sq.	Sq. Root	Cube	Cube Root	No.	Sq.	Sq. Root	Cube	Cube Root
1	1	1.000	1	1.000	51	2,601	7.141	132,651	3.708
2	4	1.414	8	1.260	52	2,704	7.211	140,608	3.732
3	9	1.732	27	1.442	53	2,809	7.280	148,877	3.756
4	16	2.000	64	1.587	54	2,916	7.348	157,464	3.780
5	25	2.236	125	1.710	55	3,025	7.416	166,375	3.803
6	36	2.449	216	1.817	56	3,136	7.483	175,616	3.826
7	49	2.646	343	1.913	57	3,249	7.550	185,193	3.848
8	64	2.828	512	2.000	58	3,364	7.616	195,112	3.871
9	81	3.000	729	2.080	59	3,481	7.681	205,379	3.893
10	100	3.162	1,000	2.154	60	3,600	7.746	216,000	3.915
11	121	3.317	1,331	2.224	61	3,721	7.810	226,981	3.936
12	144	3.464	1,728	2.289	62	3,844	7.874	238,328	3.958
13	169	3.606	2,197	2.351	63	3,969	7.937	250,047	3.979
14	196	3.742	2,744	2.410	64	4,096	8.000	262,144	4.000
15	225	3.873	3,375	2.466	65	4,225	8.062	274,625	4.021
16	256	4.000	4,096	2.520	66	4,356	8.124	287,496	4.041
17	289	4.123	4,913	2.571	67	4,489	8.185	300,763	4.062
18	324	4.243	5,832	2.621	68	4,624	8.246	314,432	4.082
19	361	4.359	6,859	2.668	69	4,761	8.307	328,509	4.102
20	400	4.472	8,000	2.714	70	4,900	8.367	343,000	4.121
21	441	4.583	9,261	2.759	71	5,041	8.426	357,911	4.141
22	484	4.690	10,648	2.802	72	5,184	8.485	373,248	4.160
23	529	4.796	12,167	2.844	73	5,329	8.544	389,017	4.179
24	576	4.899	13,824	2.884	74	5,476	8.602	405,224	4.198
25	625	5.000	15,625	2.924	75	5,625	8.660	421,875	4.217
26	676	5.099	17,576	2.962	76	5,776	8.718	438,976	4.236
27	729	5.196	19,683	3.000	77	5,929	8.775	456,533	4.254
28	784	5.291	21,952	3.037	78	6,084	8.832	474,552	4.273
29	841	5.385	24,389	3.072	79	6,241	8.888	493,039	4.291
30	900	5.477	27,000	3.107	80	6,400	8.944	512,000	4.309
31	961	5.568	29,791	3.141	81	6,561	9.000	531,441	4.327
32	1,024	5.657	32,768	3.175	82	6,724	9.055	551,368	4.344
33	1,089	5.745	35,937	3.208	83	6,889	9.110	571,787	4.362
34	1,156	5.831	39,304	3.240	84	7,056	9.165	592,704	4.380
35	1,225	5.916	42,875	3.271	85	7,225	9.220	614,125	4.397
36	1,296	6.000	46,656	3.302	86	7,396	9.274	636,056	4.414
37	1,369	6.083	50,653	3.332	87	7,569	9.327	658,503	4.431
38	1,444	6.164	54,872	3.362	88	7,744	9.381	681,472	4.448
39	1,521	6.245	59,319	3.391	89	7,921	9.434	704,969	4.465
40	1,600	6.325	64,000	3.420	90	8,100	9.487	729,000	4.481
41	1,681	6.403	68,921	3.448	91	8,281	9.539	753,571	4.498
42	1,764	6.481	74,088	3.476	92	8,464	9.592	778,688	4.514
43	1,849	6.557	79,507	3.503	93	8,649	9.644	804,357	4.531
44	1,936	6.633	85,184	3.530	94	8,836	9.695	830,584	4.547
45	2,025	6.708	91,125	3.557	95	9,025	9.747	857,375	4.563
46	2,116	6.782	97,336	3.583	96	9,216	9.798	884,736	4.579
47	2,209	6.856	103,823	3.609	97	9,409	9.849	912,673	4.595
48	2,304	6.928	110,592	3.634	98	9,604	9.899	941,192	4.610
49	2,401	7.000	117,649	3.659	99	9,801	9.950	970,299	4.626
50	2,500	7.071	125,000	3.684	100	10,000	10.000	1,000,000	4.642

Appendix: Answers to Problems

In the event that your answers to the computed values of the statistical tests approximate, but do not precisely equal, those given below, you should first consider the "number of places" used in the different computations in order to understand discrepancies.

Chapter 5

1. With 26 *df*, a *t* of 2.14 is significant beyond the .05 level. Hence the null hypothesis may be rejected.

2. With 30 *df*, a *t* of 2.20 is significant beyond the .05 level. Since this was the criterion set for rejecting the null hypothesis, and since the direction of the means is that specified by the empirical hypothesis, it may be concluded that the empirical hypothesis was confirmed—that the independent variable influenced the dependent variable.

3. With 13 *df*, the computed *t* of 4.55 is significant beyond the 1% level. Since the group that received the tranquilizer had the lesser mean psychotic tendency it may be concluded that the drug produces the advertised effect.

4. The computed *t* of .52 is not significant. Since the experimenter could not reject his null hypothesis, he failed to confirm his empirical hypothesis.

5. His suspicion is not confirmed—the computed *t* is .15.

Chapter 6

1. The confounding in this study is especially atrocious. The subjects in the two groups undoubtedly differ in a large number of respects other than type of method. For instance, there may be differences in intelligence, opportunity to study, socio-economic level, as well as differences in reading proficiency prior to learning by either method, and certainly there were different teachers. The proper approach would be to randomly assign subjects from the same class in a given school to two groups, and then to randomly determine which group is taught by each method, both groups being taught by the same instructor.

2. The characteristics of the individual tanks and targets are confounded with the independent variable. It may be that one tank gun is more accurate than the other, and that one set of targets is easier to hit than the other. To control these variables one might have all subjects fire from the same tank (continually checking the calibration of the gun) on the same set of targets. Or half of the subjects from each group could fire from each tank onto each set of targets.

3. The conclusion reached in this study is limited to the effects of class from which the children came. Undoubtedly these classes differ in a number of respects, among which is age at which they are toilet trained. The dependent variable results may thus be due to some other differential experience of the groups such as amount of social stimulation, or amount of money spent on family needs. The obvious, but difficult, way to conduct this experiment in order to establish a causal relation would be to randomly select a group of children, randomly assign them to two groups, and then randomly determine the age at which each group is toilet trained.

4. The control group should also be operated on, except that the thalamus should not be impaired. It could be that some structure other than the thalamus is disturbed during the operation, and this other structure may be responsible for the "missing" behavior.

5. There may be other reasons for not reporting an emotionally loaded word than that it is not perceived. For instance "sex" may actually be perceived by a subject, but he waits until he is absolutely sure that that is the word, possibly saving himself from a "social blunder." In addition, the frequency with which the loaded and neutral words are used in everyday life undoubtedly differ, thus affecting the threshold for recognition of the words. A better approach would be to start with a number of words that are emotionally neutral (or with nonsense syllables), and make some of them emotionally loaded (such as associating an electric shock with them). The loaded and neutral words should be equated for frequency of use.

Chapter 8

1. With 7 df the computed t of 2.58 is significant beyond the 5% level. Hence the null hypothesis may be rejected. However the subjects who used the Eastern Grip had a higher mean score, from which we can conclude that the empirical hypothesis is not confirmed.

2. The computed t of 2.98 with 7 df is significant beyond the .05 level. Since the experimental group had the higher mean score, the empirical hypothesis is confirmed.

3. With 19 *df*, the computed *t* of 7.02 is significant beyond the .02 level. Since the group that used the training aid had the higher mean score, we may conclude that the training aid facilitated map reading proficiency.

Chapter 9

1. $R_2 = 1.04$ and $R_3 = 1.09$.

Mean Scores for Groups

English Majors	Art Majors	Chemistry Majors
2.17	5.50	9.33

All groups are significantly different from each other.

2. $R_2 = 2.42$, $R_3 = 2.54$, and $R_4 = 2.62$.

Mean Scores for Groups

III	II	I	IV
2.00	2.38	6.25	7.12

Groups II and III are both significantly different from Groups I and IV. It might be added that, for greatest proficiency, these fictitious data indicate that considerable practice or extremely little practice are most beneficial (a U-shaped curve).

3. $R_2 = 2.28$, $R_3 = 2.40$, $R_4 = 2.47$, and $R_5 = 2.53$.

Mean Scores for Groups

1	2	3	4	5
3.45	3.82	4.18	6.64	8.45

The order of means increases systematically with the independent variable. The higher two means are significantly superior to the lower three means. In general, the hypothesis was confirmed. It would have been more desirable, however, to have obtained a significant difference between each of the groups, a goal that might have been achieved had a larger number of subjects per group been studied.

4. R_2 for the comparison between Groups B and C is 7.36, between Groups A and B is 8.33, and R_3 for the comparison between Groups A and C is 8.35.

Mean Scores for Groups Taught by:

Method C	Method B	Method A
27.23	29.90	46.62

Method A is to be preferred since it led to significantly greater proficiency than did the other two methods.

Chapter 10

1.

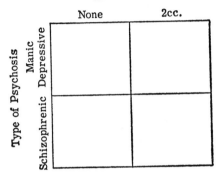

Amount of Drug
Administered

2.   Analysis of Variance

Source of Variation	Sum of Squares	df	Mean Square	F
Over-all Between	(91.13)	(3)		
Between Drugs	89.28	1	89.28	39.33
Between Psychoses	1.28	1	1.28	.56
D × I	.57	1	.57	.25
Within Groups	54.58	24	2.27	
Total	145.71	27		

Since the *F* for "between drugs" is significant variation of this independent variable is effective. The mean score for the subjects who received drugs is higher than that for those who did not receive drugs. Hence we may conclude that administration of the drug led to an increase in normality. The lack of significant Fs for the "between psychoses" and interaction sources of variation indicates that there is no difference in normality as a function of type of psychosis, nor that there is an interaction between the variables.

3.

Amount of Drug Administered

	None	2cc.	4cc.
Paranoid			
Manic Depressive			
Schizophrenic			

Type of Psychosis

4.

Amount of Drug Administered

	None	2cc.	4cc.	6cc.
Paranoid				
Manic Depressive				
Schizophrenic				
Normal				

Type of Subject

5.

Type of Brand

    Old Zincs   Archdukes

6.   Analysis of Variance

Source of Variation	Sum of Squares	df	Mean Square	F
Over-all Between	(275.88)	(3)		
Between Brands	0.03	1	0.03	0.02
Between Filters	0.23	1	0.23	0.12
B × F	275.62	1	275.62	144.30
Within Groups	68.90	36	1.91	
Total	344.78	39		

Since neither variation of brands nor filters resulted in significant differences, we may conclude that variation of these variables, considered by themselves, did not affect steadiness. However the interaction was highly significant. From this we may conclude that whether or not brand affects steadiness depends on whether or not a filter was used—that smoking Old Zincs with a filter leads to greater steadiness than does smoking Archdukes with a filter, but that smoking Archdukes without a filter leads to greater steadiness than smoking Old Zincs without a filter. It would appear that putting a filter on Archdukes decreases steadiness, but putting a filter on Old Zincs increases steadiness. In fact, Archdukes without a filter leads to about the same amount of steadiness as Old Zincs with a filter, as a diagram of the interaction would show.

7.   *Analysis of Variance*

Source of Variation	Sum of Squares	df	Mean Square	F
Over-all Between	(80.68)	(3)	26.89	
Between Opium	30.04	1	30.04	27.81
Between Marihuana	48.89	1	48.89	45.27
O × M	1.75	1	1.75	1.62
Within Groups	26.00	24	1.08	
Total	106.68	27		

Since the F's for "between Opium" and "between Marihuana" are both significant, we can conclude that smoking both of them lead to halucinatory activity, and that there is no interaction between these two variables since this latter source of variation is not significant.

Index

305